Children's Health and the Peril
of Climate Change

CHILDREN'S HEALTH
AND THE PERIL
OF CLIMATE CHANGE

Frederica Perera

OXFORD
UNIVERSITY PRESS

OXFORD
UNIVERSITY PRESS

Oxford University Press is a department of the University of Oxford. It furthers the University's objective of excellence in research, scholarship, and education by publishing worldwide. Oxford is a registered trade mark of Oxford University Press in the UK and certain other countries.

Published in the United States of America by Oxford University Press
198 Madison Avenue, New York, NY 10016, United States of America.

Library of Congress Cataloging-in-Publication Data
Names: Perera, Frederica P., author.
Title: Children's health and the peril of climate change / by Frederica Perera.
Description: New York, NY : Oxford University Press, [2022] |
Includes bibliographical references and index.
Identifiers: LCCN 2022018918 (print) | LCCN 2022018919 (ebook) |
ISBN 9780197588161 (hardback) | ISBN 9780197588178 (epub) |
ISBN 9780197588192
Subjects: MESH: Child Health | Climate Change | Fossil Fuels—adverse effects
Classification: LCC RJ61 (print) | LCC RJ61 (ebook) | NLM WS 440 |
DDC 618.92—dc23/eng/20220606
LC record available at https://lccn.loc.gov/2022018918
LC ebook record available at https://lccn.loc.gov/2022018919

DOI: 10.1093/oso/9780197588161.001.0001

9 8 7 6 5 4 3 2 1

Printed by Sheridan Books, Inc., United States of America

CONTENTS

PREFACE

Evolving as a scientist, mother, and now grandmother, I have come to understand the urgent need to recognize the full measure of harm being inflicted on our children by fossil fuel–related climate change and air pollution. The state of the climate has been described as "code red for humanity." It is especially "code red" for children, who bear the brunt. I intend to show the scope of both the challenge and the opportunity, using the solutions now available, to provide a healthy and equitable future for them. I hope to engage everyone who cares about the health of our children to become involved because I believe that by working together we can meet the challenge.

My own evolution can be told, in part, through a bit of background on my own academic history. As a graduate student at Columbia, I was fascinated by the work being done at the laboratories of Columbia and the National Cancer Institute to understand the mechanism by which a pervasive and toxic class of air pollutants—polycyclic aromatic hydrocarbons (PAHs)—could be causing disease. The team had devised a laboratory method to measure the amount of PAH latched on to DNA, damaging the genetic material. They were using that "biomarker" in their experimental research on cancer. I wondered whether the PAH–DNA "adduct," as it was called, could be detected in human blood and tissue and, if so, could serve as a marker of exposure and potential risk. For my research, I sought a pristine human tissue, one that did not contain any PAH–DNA adducts, for comparison with samples from adults. Wouldn't placental tissue, newborn cord blood, or placenta be such a "control" sample, unmarked by any environmental exposure? To my dismay, these supposedly pristine human samples were found to contain measurable levels of PAH–DNA adducts, destroying that myth and shocking me into a new research path. I needed to find out more about the harm from exposure to air pollutants and other toxic chemicals during fetal development by measuring such biomarkers in cord blood and studying their relationship to the health of the children as they developed. Thus began a long journey.

Over the past decades, my research with my colleagues has highlighted the exquisite vulnerability of the developing fetus to environmental insults and has been instrumental in recognizing the prenatal window of susceptibility. It has uncovered associations between early-life exposures and multiple adverse health effects, revealed the health benefits of interventions to reduce toxic exposures, and helped to spur policy change. Thus, research that began as an abstract theoretical problem opened hopeful new avenues for action to protect children.

For many people climate change has either been an abstraction with no apparent relevance to their lives or a dire prediction they have chosen to look away from in order to continue their lives as usual. Many have been blocked by the perception that the situation is hopeless and there is little that can be done to avoid the worst. The task seems too big and the required personal sacrifices unjustified. As a result, the energy and power of a large percentage of the population have been untapped in the battle to stabilize the climate and avoid a public health catastrophe for our children. My goal is to make that connection and to motivate parents, other family members, policymakers, and the wider public to act on behalf of the health of all children and their future.

To build awareness of the sheer magnitude of the environmental impacts of fossil fuel and the health burden now being inflicted on children everywhere in the world, I lay out the science regarding the unique susceptibility of the young and show that, while all children are at risk, poverty and racism greatly increase the impacts of climate change and pollution. I describe the full scope of the harm of climate change and air pollution to children's physical health, brain development, and mental health, harm beginning even before they are born.

I then provide facts and narratives about real-world solutions to counter denial, despair, and paralysis with hope and purpose. I describe the positive and growing power and voice of Indigenous and environmental justice groups, youth, and spiritual leaders who are motivating action by governments, communities, and individuals. I give examples of success stories from policies already in place around the world that have generated large health benefits to children, with substantial economic savings. I describe the many solutions now at hand, including actions that individuals can take in their own lives.

As I have witnessed the growing threat of climate catastrophe, unmet by action on the scale needed, I have come to realize the need to broaden the discussion of climate change to include all of the impacts of fossil fuel dependence, including the myriad health effects from toxic air pollutants emitted along with greenhouse gases. I hope that, when hearing the words "climate change," readers will not only picture melting icebergs and stranded polar bears, but also malnourished and stunted children in Somalia; families and children fleeing forest fires raging in Australia, Canada, and California; flooding in Germany and South Texas; and parents and children sheltering from extreme heat in India and Florida. That they will also see asthmatic children using inhalers, tiny babies born too soon, and children struggling to learn because these are all effects of fossil fuel emissions—and they are avoidable.

It is my hope that readers will become active participants in the climate fight when they clearly see the toll of inaction on our children and, conversely, the enormous benefits in terms of avoided deaths and illness and a sustainable future for them if we act. And that we will act to protect our children's health and future, avoiding the fate of being the first generation in human history to knowingly leave our children in peril.

I would like to acknowledge and thank the many friends and colleagues who have inspired me to write the book: Wendy Neu, Frances Beinecke, Sarah Chasis, Cynthia Ryan, my colleagues from Columbia including the late I. Bernard Weinstein, Julie Herbstman, Deliang Tang, Robin Whyatt, Virginia Rauh, Rachel Miller, Andrew Rundle, and Jack Mayer, and the late Wieslaw Jedrychowski at the Jagiellonian University in Krakow. My deepest gratitude to my husband, Fritz, and to my children, grandchildren, and other family members who have encouraged me to tackle this large topic and have borne with me as I did. My grateful thanks also to Kaitlyn Coomes and Sarah Levine for help in manuscript preparation and my editor at Oxford University Press, Sarah Humphreville, for her expert guidance.

PART I

What We Know

CHAPTER 1

The Challenge of Fossil Fuel and Climate Change

INTRODUCTION

This chapter is intended to help us better understand the twin threats to children of climate change and air pollution, both largely driven by fossil fuel emissions.[1] Much of this will not be new to the casual reader of *The New York Times* or consumer of social media, but in order for us to fully appreciate the dire situation our children face, we cannot skip this summary. We will see close up the worsening state of the climate, the degradation of the air we breathe, and the frightening speed at which we have arrived at an existential crisis. I will trace the problem directly to the relentless increase in emissions generated by what Naomi Klein refers to as our "fossilized economy."[2]

In this chapter we will see how our understanding of the climate crisis has grown exponentially since Wallace Broecker at Columbia University first used the term "global warming" in a 1975 paper in *Science* magazine. By 1990, the UN Intergovernmental Panel on Climate Change (IPCC) had affirmed that human activities were substantially responsible; 30 years later, terms like "existential threat" and "code red for humanity" were being routinely used to describe the state of the climate.[3]

Every story needs a villain, and fossil fuel ably fills that role: it is the primary source of both carbon dioxide (CO_2), the most important greenhouse gas (GHG) emitted by human activity, and air pollution. It is shocking to see in black and white the staggering quantities of air pollutants, CO_2, and other GHGs released into the air and the breadth and depth of their

Children's Health and the Peril of Climate Change. Frederica Perera, Oxford University Press. © Oxford University Press 2022. DOI: 10.1093/oso/9780197588161.003.0001

impacts. Equally shocking are the trajectories of emissions, temperature, ocean acidification, disasters, fires and drought, biodiversity loss, child ill health, and inequality—all accelerating upward. Tipping points and feedback loops, some only recently recognized, prevent our knowing with certainty where these trends will take the Earth and ourselves.

The word "trajectory" means "the path followed by a body moving under the action of given forces," from the modern Latin *trajectorium*, relating to being thrown or hurled across space. In the case of climate change, the word "trajectory" implies a certain fatalism: that we and our world are being propelled toward a terrible fate by powerful forces out of our control. That is, however, not the case. The forces hurling us toward climate disaster are our built-in economic dependence on fossil fuel and our passivity due to ignorance, denial, or despair. As the youth activists point out, generations of adults had the will, ingenuity, and energy to get us into this mess. And we can get us out. In Chapter 6, "Success Stories," and Chapter 7, "Solutions Now," we will see that the predicted trajectories are certainly not inevitable and that we as a society and as individuals have the power to change them. In fact, as we confront the dire situation presented in this chapter, we will occasionally be cheered by signs of progress.

FOSSIL FUEL CONSUMPTION

In 1950, the global consumption of oil, coal, and gas began a steep ascent, increasing eight-fold between 1950 and the present. In 2021, global fossil fuel consumption, mainly by industry, transportation, and the residential sector, reached an all-time high at nearly 15 billion metric tons of fossil fuels. Close to 85 percent of global primary energy comes from coal, oil, and gas. Although oil and gas consumption continue to increase worldwide, coal consumption has dropped in many parts of the world.[4]

In the United States, the transportation and industrial sectors consume more than half of the fossil fuel–produced energy, followed by the residential and commercial sectors. Oil and natural gas are now responsible for about 70 percent of current primary energy consumption, followed by renewables, coal, and nuclear. The year 2019 was the first year that renewable energy consumption in the United States exceeded coal consumption.[5]

There is a sobering new addition to the list of consumers of fossil fuel. Faced with the inevitability of the transition to clean energy, fossil fuel producers have turned for rescue to a profitable, new market for oil and gas—the petrochemical industry. Petrochemicals are the primary components of crude oil and gas used to make plastics, pesticides, fertilizers, and a multitude of consumer products. Production trends are sharply up—so much so that petrochemicals are on track to account for more than a third of the growth in oil demand by 2030 and nearly half of that growth by 2050. Petrochemical production will likely consume an additional 56 billion cubic meters of natural gas by 2030, equivalent to about half of Canada's total natural gas consumption today. Producers are drastically expanding their footprint with huge new petrochemical plants being built in the United States, Europe, the Middle East, Asia, and Latin America.[6]

CO_2, METHANE, AND OTHER GREENHOUSE GASES

The growth trajectory of GHG emissions parallels that for fossil fuel consumption over time—not surprising since the production and combustion of fossil fuels is the largest source of CO_2, methane, and other GHGs. Scientists summarize these GHG emissions in terms of total CO_2 *equivalents*, based on their global warming potential relative to CO_2—the predominant GHG that represents 76 percent of the total.

Global annual emissions of CO_2 have grown at an alarming pace. In the 200 years between the start of the Industrial Revolution and 1950, CO_2 emissions rose steadily but then took a sharp upturn, skyrocketing to 35 billion metric tons of CO_2 by 2020. By the end of that year, a staggering 55 billion tons of CO_2 equivalents had been released by human activities. So rapid has been the growth in fossil fuel combustion that more than half of the carbon now in the atmosphere was emitted just in the past three decades.[7]

China leads the world in CO_2 emissions, having bumped the United States to number two in 2004. India, Russia, and Japan follow in order. Globally, the major sources of CO_2 emissions, in order of importance, are the burning of coal, oil, and gas for electricity and heat production, industry, and transportation. In the United States, oil is the leading source of CO_2 emissions from energy consumption, followed by natural gas and coal. Of the end-use sectors transportation emits the most CO_2, followed by power generation, industry, and buildings.[8]

There was some encouraging news in 2019. Although 2019 was a record year for total global CO_2 emissions, emissions from energy production flattened at around 33 billion tons, after 2 years of increases. The flattening was largely due to the drop in CO_2 emissions from the power sector in advanced economies as they expanded renewable sources, switched from coal to natural gas, and increased output from nuclear power. The United States saw a decline of almost 3 percent in energy-related CO_2 emissions in 2019, due mainly to a 15 percent lower draw on coal for power generation and a greater share of energy from natural gas and renewables. The European Union (EU), including the United Kingdom, did even better: energy-related CO_2 emissions dropped by 5 percent in 2019 due to increasing renewables and switching from coal to gas in the power sector. In Japan, energy-related CO_2 emissions also fell. However, these gains were countered by an increase in energy-related CO_2 emissions outside the advanced economies, with almost 80 percent of the increase coming from Asia where the demand for coal—the very dirtiest fuel—continued to expand, accounting for more than 50 percent of energy use.[9]

In a sobering development, in 2021 global energy-related carbon dioxide emissions rose to their highest recorded level, as the world economy rebounded from the Covid-19 crisis and relied heavily on coal to power that growth. For the same reasons, in 2021 United States GHG emissions in 2021 increased 6% relative to 2020, though emissions remained below 2019 levels. In addition, the Russian invasion of Ukraine in the Spring of 2022 has led many countries to turn to coal or imports of liquefied natural gas as alternative sources to Russian energy. The growth in natural gas, a source of 7.5 billion metric tons of CO_2 equivalents globally in 2000–2021, is further threatening gains from reduced coal burning in the higher-income countries. Natural gas is the fastest growing fossil fuel worldwide, with the United States leading the boom. In the United States, natural gas surpassed coal as a source of energy-related CO_2 emissions several years ago.[10]

Methane is second in importance to CO_2 as a GHG, making up 16 percent of manmade GHG emissions. Although it is much more short-lived than CO_2 (with a lifetime in the atmosphere of around 12 years), due its higher energy absorption methane has 84 times the global warming impact as CO_2 over a 20-year period. As a result, around 25 percent of current global warming can be traced to methane. Globally, human activities contribute 60 percent of total methane emissions, with fossil fuel the second largest source after agriculture. In a recent year, worldwide oil and natural gas operations emitted 82 million metric tons of methane; in the United States these operations released 13 million metric tons. There was some good news in 2021: almost three dozen countries, together accounting for 30 percent of global methane emissions, joined a pledge to cut methane emissions by 30 percent by 2030.[11]

The story of natural gas is a cautionary one. Natural gas has been touted as the ideal "bridge fuel" to a low-carbon economy because it produces half as much CO_2 as coal and 30 percent less CO_2 than oil per unit of energy delivered, as well as far less air pollution. But that story is being rewritten as growth in natural gas consumption has shot up and its full impact as a source of CO_2 and, particularly, methane has become clear. A technique called *hydraulic fracturing* or "fracking" is used to extract natural gas or oil from shale and other forms of "tight" rock by blasting water, chemicals, and sand into these formations under high pressure to crack the rock and allow the trapped gas and oil to flow to the surface. In addition to the many environmental impacts (contamination of water and air and triggering of earthquakes to name a few), the overall process from drilling to arrival at its point of use releases methane. Methane leaks may be accidental, occurring during production and transport via pipelines, but the gas is also intentionally released during controlled burns called "flaring." Then there are the accidents: over a 20-day period in 2018, a blowout at a natural gas well in Ohio released an amount of methane greater than the annual anthropogenic methane emissions from all but three EU member countries. So, natural gas is doing double duty in driving climate change through emissions of both methane and CO_2. For these reasons, natural gas has been described as "a bridge to nowhere."[12]

The petrochemical industry is a rapidly growing source of GHGs. In addition to GHG releases during the extraction and transport of fossil fuels that are the feedstock for petrochemicals, the production of petrochemicals from oil and gas requires very large amounts of energy. So energy-intensive is the process of producing petrochemical-based products that the industry is already one of the major sources of CO_2. Global emissions of CO_2 linked to plastic alone, now almost 900 million tons of CO_2 equivalents annually, could reach 1.3 billion tons by 2030—as much as the emissions from 300 coal-fired power plants. If output grows as planned, by 2050 GHG emissions from the plastic lifecycle could reach more than 56 billion metric tons, equivalent to 615 coal-fired power plants, making it impossible to limit global temperature rise to 1.5°C (2.7°F).[13]

It is shocking to learn that relatively few companies have been responsible for the majority of GHG emissions. Since 1988—the year the UN IPCC was established to track and report on the science of climate change—more than half of global industrial GHG emissions can be traced to just 25 corporate and state-owned entities.[14]

LOSS OF CARBON "SINKS"

Each year, the earth's oceans, forests, plants, and soils take up billions of tons of CO_2 from the atmosphere. But since the Industrial Revolution carbon emissions from human activity have exceeded the capacity of these "natural sinks" to absorb them. Currently only half of the CO_2 released from the burning of fossil fuels every year is taken up by natural sinks; the rest is left to linger in the atmosphere for hundreds and even thousands of years. Due to the surge in emissions and its accumulation in the atmosphere, as of today the CO_2 concentration is 50 percent above pre-industrial levels.[9]

We have seen a disturbing drop in the ability of forests to act as carbon sinks. Covering 31 percent of the land area on earth, forests have played a critical role in accumulating and storing carbon dioxide that otherwise would be released to the atmosphere. Recent years have witnessed a decimation of forest cover: more than 160,000 square miles, an area roughly the size of California, were lost to deforestation in just 24 hotspots around the world between 2004 and 2017. In 2019, the tropics lost close to 30 soccer fields' worth of trees every single minute. In a recent year, loss of tropical forests from clearing or burning contributed about 4.8 billion metric tons of CO_2 (or about 8–10% of annual human of CO_2 emissions). About half of the deforestation occurs in the Amazon, the largest tropical forest on the planet, where the pace has recently surged to its highest rate in a decade. Since 1978, about 1 million square kilometers of Amazon rainforest have been destroyed by fires mostly lit for clearing, agricultural clearcutting, livestock ranching, logging, and mining (Figure 1.1). Seventeen percent of

Figure 1.1 Aerial view of deforestation in the Amazon rainforest, near Belém, Brazil.
Credit: Sue Cunningham Photographic/Alamy Stock Photo.

the Amazon has been lost in the past 50 years. Reaching its highest rate of deforestation in 15 years, the Amazon lost more than 5,000 square miles of tree cover from August 2020 to July 2021.[15]

Scientists are now warning that the Amazon is likely to become a net *emitter* of GHGs from combustion of forest biomass, decomposition of remaining plant material, and release of the carbon stored in soil. In fact, a decade-long study suggests that roughly 20 percent of the Amazon has already become a carbon source instead of a carbon sink due mainly to deforestation.[16]

The Amazon rainforest plays many other roles: it breathes out a fifth of Earth's oxygen and cycles 20 percent of the world's fresh water through its rivers, plants, soils, and air, releasing moisture that fuels rainfall in distant regions as far away as Texas. It is also the ancestral home of 1 million Indigenous people whose cultures and livelihood are threatened, along with 3 million species of insects, plants, birds, and other forms of life—one in ten of the known species on Earth.[17]

AIR POLLUTION

The combustion of fossil fuels has created a parallel crisis of air pollution. Air pollution increased steadily following the Industrial Revolution due to the burning of fossil fuels, which continues to be the world's largest contributor to air pollution. Combustion of coal, diesel, gasoline, oil, and natural gas in high- and middle-income countries, together with burning of plant and animal material in low-income countries, generates 85 percent of all airborne respirable particulate pollution (also referred to as fine particulate matter or $PM_{2.5}$). Every year about 48 million tons of respirable particles are dumped into the air by human activities. The particles are so tiny—30 times smaller than a human hair—that they escape the defenses in the upper lung and penetrate deep into the lung where they dispense the toxic chemicals carried on their surface. Fossil fuel burning is also responsible for almost all sulfur dioxide (SO_2) and nitrogen oxide (NO_x) emissions to the atmosphere, as well as large quantities of polycyclic aromatic hydrocarbons (PAH), mercury, and volatile chemicals that form ground-level ozone. All of these pollutants are harmful to children's health, including when exposure occurs while they are in the womb.[18]

Showing that regulations work, after a surge in air pollution in the preceding decades, average global air pollution exposures declined from 2010 to 2016 due to laws in North America and Europe that were implemented 30 years earlier. However, half the world's population—notably in Central and Southern Asia and Sub-Saharan Africa—is suffering increasingly higher levels of air pollution. As much as 90 percent of the total global population is exposed to air pollution above World Health Organization (WHO) guidelines for health protection. One billion children worldwide are exposed to very high levels of air pollution.[6] Air pollution affects people in every country in the world, but low- and middle-income countries and disadvantaged populations everywhere bear the greatest burden.[19]

In the United States, the current picture is mixed: after a 24 percent decline from 2009 to 2016, annual average fine particulate matter increased by about 6 percent between 2016 and 2018, more than offsetting the benefit of the reduction in coal burning. In the next year $PM_{2.5}$ emissions climbed to 5.7 million tons. This unfortunate reversal is attributable to increases in use of natural gas, increased driving, and more wildfires, as well as the rollback by the Trump Administration of Environmental Protection Agency (EPA) regulations (which, happily, the current Biden Administration is restoring).[20]

THE CONSEQUENCES OF CLIMATE CHANGE

Warming

Largely as a result of fossil fuel emissions, the Earth's average surface air temperature has increased by 1.1°C (2°F) compared to pre-industrial times. Figure 1.2 shows the steeply rising curve of global warming in the past

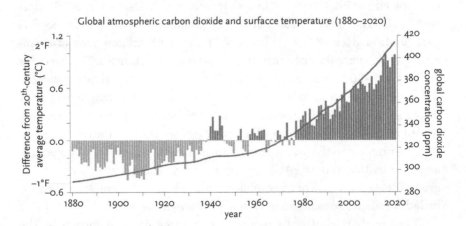

Figure 1.2 This graph shows the sharp rise in the yearly average temperature after 1980 (values on the left axis). The black line shows the marked upward trend in yearly atmospheric carbon dioxide (CO_2) concentrations in the past four decades (values on right axis).
Credit: Original graph by Dr. Howard Diamond (NOAA ARL) and adapted by NOAA Climate.gov.

three decades, paralleling the trend for GHG emissions from fossil fuel combustion over the same period.[21]

According to scientists at the National Aeronautics and Space Administration (NASA) Goddard Institute for Space Studies, the global average temperatures during the past 7 years have been the warmest on record, with a record high in 2020. Remarked the Institute Director Gavin Schmidt: "With these trends, and as the human impact on the climate increases, we have to expect that records will continue to be broken." We also have to be aware that global averages only take us so far in grasping the reality of global warming. Warming greater than the global annual average is being experienced in many land regions and seasons: the Arctic is experiencing warming two to three times higher than the global average, causing thawing of the frozen arctic soil (permafrost) and releasing CO_2 that has been stored within it for thousands of years.[22]

To address climate change on a global scale, in 2015, leaders from 196 countries signed an international treaty known as the Paris Agreement. This legally binding agreement aimed to keep the global temperature rise during this century below 2°C above pre-industrial levels and to pursue efforts to limit the temperature increase to 1.5°C. Former US President Trump abandoned the treaty; but in February 2021, under the new Biden presidency, the United States officially rejoined the Paris Agreement. In 2015, many politicians and scientists cheered the event as a historic global turning point for change. But some, including public health and climate change expert Andy Haines, noted at the time that, even if the reductions in GHG emissions set in the Paris Agreement were fully implemented, by the end of the century the global average temperature would still have increased by about 3°C (5.4°F) above pre-industrial levels. Therefore, Haines concluded, deeper cuts are required if we are to have a hope of keeping the increase to well below 2°C (3.6°F). After attending the 2015 Paris conference and studying the details of the agreement, I shared his concern and saw the Agreement as just a first step.[23]

Haines's thesis is proving more accurate every year. Recent research has just ruled out the previous low-end estimates of warming, forecasting instead that, at the current rate of CO_2 emissions and without aggressive action on the part of governments, the likely range of warming over the next 50 years is between 2.6°C (4.7°F) and 4.1°C (7.3°F), with a best estimate of about 3°C above preindustrial levels. Such a level of average warming would cause permanent damage to ecosystems, devastating heat waves, sea level rise, and a host of other problems that will affect hundreds of millions of people in coastal cities and communities around the world. In other words, continuing on our current path will result in climate catastrophe.[24]

A Special Report on Global Warming in 2018 by the IPCC warned that even limiting global warming to 1.5°C would not fully spare human health, ecosystems, livelihoods, and the economy from challenging impacts. But it concluded that, compared with 1.5°C, an upper limit of 2°C would significantly increase those impacts: we would see more drought, coastal flooding, extreme heat, food scarcity, and climate-related poverty for hundreds of millions of people. There would be drastic loss of species with an increase of 2°C—more than 99 percent of corals would be lost—and irreversible instabilities could be triggered in Antarctica and the Greenland ice sheet resulting in a multi-meter rise in sea level. James Hansen, the NASA scientist who was instrumental in raising awareness of climate change, noted that both 1.5°C and 2°C would take humanity into "uncharted and dangerous territory because they were both well above the range during the Holocene-era (about 12,000 years ago to the present) in which human civilization developed." But, he added, there was a huge difference between the two: "1.5°C gives young people and the next generation a fighting chance of getting back to the Holocene or close to it."[25]

In 2018, the IPCC concluded that, for global warming to be limited to 1.5°C, global net human-caused emissions of CO_2 would need to fall by about 45 percent from 2010 levels by 2030, reaching net zero around 2050. Limiting global warming to 1.5°C is possible, the report stated, but will require deep emissions reductions and unprecedented changes in energy, land, urban infrastructure, and industrial systems. Most sobering of all, the IPCC said that we have only a dozen years for global warming to be kept to a maximum of 1.5°C. The needed drastic reduction of emissions by 2030 and the 12-year window for meaningful action attracted wide public attention around the world and further energized youth and other activist groups. In August 2021, the IPCC issued another report, describing adverse outcomes of climate change as unprecedented and likely to breach 1.5°C above pre-industrial temperatures in the near term. UN Secretary-General António Guterres called the report "code red for humanity." However, he ended on the semi-positive note that aggressive efforts by top-emitting countries could avoid the worst climate outcomes.[26]

The 26th annual Conference of Parties (COP) of the UN Framework Convention on Climate Change (COP26) in Glasgow, in November 2021, set some new long-term targets for reduction in CO_2 emissions. The most optimistic analysis by the International Energy Agency (IEA) concluded that these new targets plus all those agrees on previously would be enough to hold the rise in global temperatures to 1.8°C (3.2°F) by the end of the century—but that is assuming all the targets are met in full and on time. Moreover, 1.8°C fails to meet the goal of limiting global warming to 1.5°C.

The authors further noted that, taken together, the climate pledges made so far leave a 70 percent gap in the amount of emissions reductions needed by 2030 to keep 1.5°C within reach.[27]

According to Bill McKibben, founder of the grassroots climate campaign 350.org and the Schumann Distinguished Scholar at Middlebury College, the solution is to keep 80 percent of the fossil-fuel reserves that we know about underground. "If we don't—if we dig up the coal and oil and gas and burn them—we will overwhelm the planet's physical systems, heating the Earth far past the red lines drawn by scientists and governments." The reserves that must remain untouched, he said, are "carbon bombs" like oil in the Arctic and the tar sands of Canada; coal in Western Australia, Indonesia, China, and the Powder River Basin; and natural gas in Eastern Europe.[28]

Severe Winter Weather Events

Over the past two to three decades, as the Arctic has warmed, a seesaw winter temperature pattern known as the *warm-Arctic/cold-continents pattern* has emerged in parts of North America and Eurasia. During mid- to late-winter of recent decades, when the Arctic warming trend is greatest, the northeastern United States has experienced more frequent cold spells and heavy snows. Climate change may be fueling this seemingly counterintuitive pattern by destabilizing the band of strong westerly winds above the North Pole known as the polar vortex, thus causing the polar jet stream below to waver and allowing cold Arctic air to move south. In addition, climate change has warmed the Atlantic Ocean, providing more water to winter storms and intensifying snowfall.[29]

OCEANS

For eons the oceans, which now cover more than 70 percent of the earth's surface, have been faithfully performing their role as the planet's most important carbon sink, taking up CO_2 through photosynthesis by phytoplankton or by dissolving the gas in water. Since the beginning of the Industrial Revolution the oceans have absorbed almost 30 percent of global CO_2 emissions. But both the service they have performed and climate change itself have taken an enormous toll on them.[30]

Ocean warming and ocean acidification have been on the same dangerous upward and accelerating trajectory we have seen with other fossil fuel impacts. The oceans store more than 90 percent of the excess heat on the planet. Measuring ocean heat content is therefore one of the best ways to quantify the rate of global warming. In 2019, an international team of scientists from institutes across the world analyzed data on measurements of ocean heat content taken over the past 60 years. Their findings clearly show a dramatic escalation in ocean temperature in the past several decades: from 1955 to 1986, the rise had been pretty steady, then it accelerated between 1987 and 2019 to a level 450 percent greater than in the previous period. The 5 years leading up to the all-time high in 2019 were the 5 warmest years in the ocean since the 1950s. Said the lead author, Lijing Cheng from the Chinese Academy of Sciences, "The amount of heat we have put in the world's oceans in the past 25 years equals 3.6 billion Hiroshima atom-bomb explosions." Put another way, that is equal to dropping about four Hiroshima bombs into the oceans every second over the past 25 years. Due to the accelerated rate of ocean heating, study coauthor John Abraham added that, "We are now at five to six Hiroshima bombs of heat each second."[31]

Warming of the oceans has had serious consequences. For example, higher ocean temperatures have affected coral, fish, and other sea life— sometimes fatally—because less oxygen is dissolved in the water at higher temperatures. In the case of the North Atlantic right whales, warming water has caused a sharp decline in the crustaceans within their traditional feeding grounds. Another effect of higher ocean temperatures is that more moisture is transferred to the air by evaporation, fueling heavy rains and intense tropical storms, including hurricanes in the Atlantic. Sea levels have risen through thermal expansion of water and melting of ice, causing widespread coastal flooding.

Warming is not the only problem for the oceans. In the past 200 years alone, ocean water has become 30 percent more acidic, faster than any known change in ocean chemistry in the past 50 million years. According to the National Oceanic and Atmospheric Administration (NOAA), based on scenarios of business-as-usual emissions, by 2100, the surface waters of the ocean could have acidity levels nearly 150 percent higher than during the Industrial Revolution. The oceans have not experienced this level of acidity for more than 20 million years.[32]

These dramatic changes in ocean chemistry are having cumulative impacts on life in the ocean. Ocean acidification is acting in concert with rising water temperatures and lower levels of dissolved oxygen to threaten more and more marine species. Increasing ocean acidification has already negatively affected marine life by affecting fishes' ability to navigate and find food by using their sense of smell and by interfering with the ability of oysters, corals, and plankton to extract calcium from the water to build their shells and skeletons. Although many species are able to adapt, acidification has already caused oyster die-offs in the US Pacific Northwest and dissolution of the shells of pteropods (tiny sea snails) in the Southern Ocean. Already stressed by warmer waters, corals in lower pH waters build thinner skeletons that are more vulnerable to damage from pounding waves and erosion by organisms that drill into or devour coral. The World Resources Institute has estimated that, unless steps are taken to reduce the emission of GHGs, by 2030 warming and acidification will threaten 90 percent of coral reefs and virtually all by 2050. This would be disastrous: coral reefs support as much as a quarter of marine life and supply food, income, and protection for more than half a billion people. The net economic value of the world's coral reefs is estimated to be tens of billions of US dollars per year. The cultural importance of these unique ecosystems to Indigenous people around the world cannot be overestimated.[33]

CLIMATE-RELATED SEVERE FLOODS, STORMS, FIRES, DROUGHT

Higher temperatures contribute to disasters in several ways. At higher temperatures, evaporation from oceans and other surface waters speeds up; warm air holds more water; more rain is produced; and resultant severe rain events ("rain bombs") lead to river flooding and higher storm surges that put coastal areas under water. Warmer seawater provides more heat energy that feeds hurricanes and other tropical cyclones, leading to bigger and stronger storms. Rising sea level has resulted from the melting of glaciers and ice sheets and the expanded volume of the ocean due to higher temperatures, causing more coastal flooding.

It is not surprising, therefore, that the frequency of climate-related "natural disasters" (now more aptly termed "unnatural disasters") has risen dramatically over the past several decades. According to the UN, climate-related disasters, mainly floods and storms, increased by more than 80 percent in the past 20 years compared to the previous two decades. Hurricanes have become more intense and destructive: a 25–30 percent global increase in the intensity of hurricanes since the 1970s has been attributed to human-caused global warming. Rain bombs have caused flash floods in rivers around the world from Israel, Jordan, and Egypt to the United States, Europe, Australia, and China. In the United States, 9 of the top 10 years for extreme 1-day precipitation events have occurred since 1996. Illustrating in an apocalyptic fashion the limitations of adaptation to storms made worse by climate change, in August 2021, the remnants of Hurricane Ida caused the largest series of flash flood emergencies ever issued by the National Weather Service, resulting in dozens of fatalities from Pennsylvania to southern New England.[34]

Coastal flooding has increased sharply both globally and in the United States, where the frequency of flooding in some cities along the Gulf and Atlantic coasts has grown five-fold since 2000. Worse, much worse, is predicted. The rate of sea level rise has been accelerating: global average sea level has risen about 8–9 inches since 1880, with about a third of that increase during the past 25 years. NOAA scientists predict that global sea level is very likely to rise at least 12 inches above 2000 levels by 2100 even on a low-emissions pathway. If we stay on the highest GHG emission pathway, a sea level as much as 8 feet above 2000 levels by 2100 cannot be ruled out. Should the Greenland ice sheet thaw in its entirety, it would add a staggering 24 feet to the height of global seas. The thaw of the West Antarctic Ice Sheet is also accelerating, which would add yet another 10 feet. The neighboring Thwaite's Glacier, comparable in size to Florida, has been fracturing and threatens to melt. If the entire glacier were to melt, global sea levels would rise about 2 feet. Even if the global temperature rise stays below 2°C, by 2050, 800 million people living in 570 low-lying coastal cities around the world will be at risk from rising seas and storm surges. These cities include New York City, Miami, Osaka, Jakarta, Bangkok, Lagos, Manila, Dhaka, and Shanghai.[35]

Wildfires are a major source of GHGs and toxic air pollutants. In Siberia, the United States, and Turkey in 2021, fires made worse by climate change emitted a record 1.76 billion tons of stored carbon into the atmosphere. During a large fire event, wildfire smoke can account for 25 percent of dangerous air pollution in the United States. Rising temperatures and drought linked to climate change have made forests drier and vegetation more flammable. Greater flammability combined with longer drying seasons and shifts in rainfall patterns have contributed to a dramatic increase in forest and wildland fires in some regions, especially the more intense fires. The year 2020 saw an unprecedented number of unusually severe wildfires around the world, notably in Australia, parts of the Arctic, the Amazon, Central Asia, and the United States. Globally, the total number of square kilometers burned each year has dropped since 2003, but that was largely due to clearing of forests and development in grasslands and savannahs.

In the western United States, we have seen a sharp increase in the intensity and spread of fires because of climate change. In July 2021, toxic smoke from more than 100 fires in the US West Coast and Canada reached East Coast cities, including New York City, where the sun glowed red and the skies were hazy for several days due to light scattering through the fine particles in smoke.[36]

In February 2022, noting that the "heating of the planet is turning landscapes into tinderboxes," a UN report sounded a loud alarm. The team of researchers estimated that, primarily because of climate change, the risk of highly devastating fires worldwide could increase by 30 percent by 2050 and 50 percent by the end of the century.[37]

Climate change has contributed to droughts that have increased worldwide in frequency, duration, and intensity at an unprecedented rate. Although climate change has increased precipitation in some areas, in other regions it has led to sharply reduced rainfall. In others, it has shifted the pattern of rainfall toward short, heavy rains, most of which runs off without soaking the ground. To make matters worse, higher temperatures have triggered earlier snow melt and drying, causing increased evaporation from soil and vegetation. The result: longer lasting and more intense drought. In the past 40 years, the percentage of the planet affected by drought has more than doubled, affecting more people worldwide than any other natural hazard. Regions all around the world from California, to the Eastern Mediterranean, East Africa, South Africa, and Australia have been afflicted by severe droughts in recent years. In the American Southwest, the period from 2000 to 2021 was the driest 22-year period since 800 AD.[38]

The human cost of extreme weather events is incalculable, poorly reflected in economic measures of the physical damage and numbers of people affected. Nevertheless, those numbers are staggering. For example, the total cost of damages from 300 major disasters (each costing $1 billion or more) since 1980 in the United States exceeds $2 trillion. In just the first 9 months of 2021, total damages from 18 major weather disasters in the United States amounted to almost $105 billion.[39]

In February 2022, the IPCC urgently warned that countries are not doing anywhere near enough to protect against the disasters to come. Among their predictions: if average warming exceeds 1.5°C, which is likely in the next several decades, there could be a 20 percent increase in the number of people globally who are exposed to severe coastal flooding; at 2°C of warming, as many as 3 billion people could face water scarcity due to drought; and the amount of land consumed by wildfires could increase by more than a third.[40]

BIODIVERSITY LOSS

In the 1980s, Stanford biologist Paul Ehrlich made the analogy that losing species in an ecosystem is like progressively popping off rivets from the wings of an airplane: the plane may capably fly on for a while, but eventually you'll have removed too many rivets and the plane will crash. Loss of plants and animal species has been accelerating at a rate faster than at any time in human history. Climate change has been one of the drivers of this decline and is growing in importance. A seminal paper published in the journal *Nature* in 2004 by Chris Thomas, then at the University of York, and his colleagues estimated that by 2050 climate change over the past 30 years will place from 15 to 37 percent of all species they studied at risk of extinction. They predicted that more than a million species could be threatened with extinction as a result of climate change and global warming, a conclusion that was widely cited in the media. Despite the uncertainties in their predictions, the analysis concluded that anthropogenic climate warming not only ranks up there beside other recognized threats to global biodiversity but is likely to be the greatest threat in many, if not most, regions.[41]

Fifteen years later, an intergovernmental report compiled by hundreds of experts from all regions of the world analyzed more than 15,000 scientific publications as well as a large body of Indigenous and local knowledge. The report identified climate change as one of the main drivers of species loss, acting in concert with changes in land and sea use, direct exploitation of organisms, pollution, and invasion of alien species. It noted the sharp declines in species to date: approximately half the live coral cover on coral reefs gone since the 1870s, with accelerating losses in recent decades, and a 20 percent drop in the average abundance of native species in most major terrestrial ecosystems. Their conclusions: extinctions have already occurred at a rate at least tens to hundreds of times higher than the average rate over the past 10 million years; the rate is accelerating; and approximately 25 percent of species are already threatened with extinction in most animal and plant groups studied. They confirmed the earlier estimate of the extent of the unfolding catastrophe: 1 million plant and animal species on the planet are at risk of extinction, many within decades, because of human activities; at risk are a half-million land-based species and one-third of all marine mammals and corals.[42]

A year later, the World Wildlife Fund and the Zoological Society of London issued a report that drew on wildlife monitoring of more than 4,300 different vertebrate species—mammals, fish, birds, and amphibians—from around the world. The dire finding: population sizes for those monitored species has declined by an average of 68 percent in less than 50 years, from 1970 to 2016.[43]

In 2014, science writer for the *New Yorker* magazine, Elizabeth Kolbert, titled her brilliant book *The Sixth Extinction*; the fifth having occurred about 66 million years ago, when a giant asteroid hit the earth causing the demise of the dinosaurs and about three-quarters of other species on earth. Among the species and ecosystems disappearing today before our eyes—or rather her eyes, since she spent time in the field and gives firsthand reports—are those in the Panamanian rainforest, the Great Barrier Reef, the Andes, Bikini Atoll, city zoos, and Kolbert's own backyard. After researching the relevant mainstream, peer-reviewed science, Kolbert estimates that between 20 and 50 percent of all living species on earth will be lost by the end of the twenty-first century if we stay on the current path. In an interview for the *New York Times*, she wryly noted, "[T]oday, you hear knowledgeable scientists say, 'We are the asteroid.'"[44]

The estimates by experts of the extent of the crisis may differ somewhat, but they are all in the same range—which is to say, horrifying. And all these experts agree that the world must take transformative action to switch from fossil fuels to a non-carbon economy as quickly as possible. We will see that biodiversity loss is on the minds—and affecting the mental health—of youth today as emblematic of the destruction of the ecosystems that support human life.

"TIPPING POINTS" AND "FEEDBACK LOOPS"

Now we come to two wildcards that shake our confidence in predictions of the future impacts of climate change. Scientists are warning us that global warming may not be happening incrementally, proceeding along a predictably rising line. Rather, that trajectory may look like a series of upward staggers as "tipping points" are passed. A tipping point may occur when a threshold in a small vital element of the climate system is exceeded, causing a significant and irreversible change in a part of the climate system. Making predictions even less certain, passing a tipping point might trigger another, causing a cascade. There is evidence that once an irreversible tipping point has been passed, a system will not revert to its original state even if the climate driver lessens or reverses: the system will have jumped to a different state. Back in 2005, James Hansen at NASA warned that "We are on the precipice of climate system tipping points beyond which there is no redemption." He was referring to the many tipping points across the Earth system, from collapsing ice sheets, melting sea ice, and thawing permafrost, to shifting monsoons and forest dieback.[45]

The disappearance of Arctic sea ice appears to be moving in the direction of such an irreversible tipping point at a rate faster than climate models have forecasted. Since 1979, the extent of Arctic ice has shrunk by 40 percent; as a result Alaska is seeing accelerated coastal erosion, loss of marine mammal habitat, and changes in the food web. A rigorous new study in

Nature reports that the planet's glaciers have lost almost 270 billion tonnes of ice every year during the period 2000–2019, accounting for a fifth of sea level rise—and the rate of melting is accelerating (Figure 1.3). (A *tonne* is a unit of mass used in the UK and Commonwealth of Nations, defined as 2,240 pounds. The unit used in the United States is *ton*, defined as 2,000 pounds.)[46]

Adding to the complexity are "feedback loops" that can either amplify or diminish the effect of a climate driver such as GHG emissions, solar irradiance, or airborne particles. Positive feedback loops occur when changing one quantity changes a second quantity, and the change in the second quantity in turn changes the first—and the cycle is repeated, over and over. An example of a strong positive feedback loop is the ice-albedo feedback in which warming causes a decrease in the area of white, light-reflecting ice caps, glaciers, and sea ice; this results in darker earth and sea surfaces; the reflection of solar energy from the Earth (the albedo) is decreased; the surface temperature of the planet is raised. This leads to more warming, there is more loss of ice, and so on. If this process isn't stopped by negative feedback, a tipping point is hit beyond which a large shift to a new state becomes inevitable.[47]

Figure 1.3 The melting of arctic sea ice is occurring at a faster rate than climate models have forecasted.
Credit: Sandra Ophorst/Shutterstock.

As if the ice-albedo feedback loop weren't concerning enough, scientists believe that the rapid melting of Arctic sea ice may start another feedback loop that rapidly melts permafrost (frozen Arctic soil), releasing vast amounts of CO_2 and methane. Permafrost, most of it in the Arctic, covers about 20 percent of the surface of the Earth and permafrost soils hold twice the amount of carbon as is now contained in the atmosphere—as much as 1,600 billion tons. As the temperature of the ground rises above freezing, soil microorganisms become active and break down the once-frozen organic carbon in the soil, releasing CO_2 and methane to the atmosphere. The increase in GHGs triggers further warming, more permafrost is lost, more GHGs are emitted, more warming and more melting permafrost follow, and the cycle continues. Today the Arctic is warming at a rate twice as fast as the rest of Earth, a pace that hasn't occurred for the past 3 million years. A recent prediction is that, if climate change continues at its current rate, half of the world's permafrost could thaw by the end of the century. New measurements indicate that Arctic permafrost could now be releasing as much as 300–600 million tons of net carbon to the atmosphere every year, indicating that the feedback loop may already be under way.[48]

Scientists are sounding the alarm about another dangerous feedback loop—this time in the Amazon Rainforest. Deforestation in the Brazilian Amazon has received a big boost under president Jair Bolsonaro: in the year between August 2019 and July 2020, deforestation surged to a 12-year high. In the past 50 years, nearly one-fifth of the rainforest has been lost to higher temperatures, clearing, and human-caused burning. Losing another fifth of the rainforest would trigger the feedback loop known as "dieback" where drying of the forest invites more wildfires and kills trees, releasing more GHGs and increasing the global temperature. This leads to more drying—and so on around the loop. Eventually this vicious cycle would cause mass tree mortality and a shift of the entire ecosystem from rainforest to savannah. The tipping point for such a collapse in the Amazon is estimated to be between 20 and 25 percent deforestation—very close to where we are now. According to Carlos Nobre, a Brazilian climate and tropical forest expert, "If you exceed the threshold, fifty to sixty per cent of the forest could be gone over three to five decades."[49]

On September 1, 2021, I checked the electronic clock on World Counts showing the hectares (1 hectare contains about 2.5 acres) of forests cut down or burned globally during the first 9 months of 2021: it was more than 18.5 million and gaining a hectare every second.[50]

There are other tipping points that could be triggered by the warming atmosphere, each hard to predict but with potentially catastrophic consequences. There is always a worst case and a range of less worrisome possibilities. But like the tipping points above, awareness that they exist should galvanize us to do everything possible to avoid them. A big one is the Atlantic Meridional Overturning Circulation (AMOC), a component of the global conveyor belt that distributes heat, energy, and nutrients around the Earth and contributes to the climate we experience today. The AMOC, also known as the Gulf Stream System, carries warm water northward along the eastern coastlines of the United States and Newfoundland before crossing the Atlantic Ocean as the North Atlantic Current to reach the British Isles and Northern Europe. The engine of the global conveyor belt is the sinking of dense, or heavy, water in the high latitudes of the North Atlantic. As the cooler water sinks, it pulls warm surface water from the Equator up north, where it chills and becomes saltier, thus becoming denser, eventually becoming heavy enough to sink to deep ocean layers and flow back to the south. Climate change affects this process by diluting the salty sea water with freshwater from increased rainfall and melting of the Greenland Ice Sheet. The less salty water is less dense, therefore lighter, and less able to sink, slowing down the very engine of this vital circulatory system.[51]

In 1997, climate scientist Wally Broecker, at Columbia University's Lamont-Doherty Earth Observatory, warned that the AMOC is the "Achilles heel" of the climate system: he cited evidence that the AMOC has repeatedly switched on and off over the course of Earth's history, with the power to flip warming periods to intense cold in the Northern Hemisphere. Research by Stefan Rahmstorf, professor of physics of the oceans at Potsdam University, has suggested that the AMOC has weakened by about 15 percent since the middle of the twentieth century and is in its weakest state in more than a thousand years. He and his team recently provided strong evidence for this disturbing conclusion by comparing archival records on temperature patterns in the Atlantic Ocean, ocean currents, and other data to reconstruct the evolution of the AMOC over the past 1,600 years. The picture of the AMOC that emerged showed a long and relatively stable period, followed by an initial weakening starting in the nineteenth century, then a second, more rapid decline in the mid-twentieth century leading to the weakest state of the AMOC occurring in recent decades. The rapid decline of this critical circulatory system is unprecedented in the past millennium. The tipping point for a shutdown is not known, but Ramsdorf thinks that the trigger lies somewhere around 3°C or 4°C (7.2°F) above pre-industrial levels. He concludes that "if we continue to drive global warming, the Gulf Stream System will weaken further—by 34 to 45 percent by 2100 according to the latest generation of climate models"; this could bring us dangerously close to the tipping point at which the flow becomes unstable. Although he does not think a complete AMOC shutdown is likely in the next 100 years, Ramsdorf and other scientists see a serious concern if we continue down the current path toward three or more degrees of warming.[52]

Even the modest slowing of 15 percent has had consequences for the patterns of ocean temperature. The rapid warming of waters off the coast of New England have contributed to the decline of the commercial cod fishery. A recurrent "cold blob" in the ocean to the south of Greenland is seen by many scientists as evidence that less warm water is now reaching this region due to the weakening of the AMOC. Worse may be to come: a further slowdown of the AMOC is predicted to cause water to pile up at the US east coast, leading to flooding of coastal cities and towns. An AMOC shutdown would leave Great Britain cooler and drier, causing widespread cessation of crop growing and farming across the country. It would also bring more extreme winter storms sweeping off the Atlantic into Europe.[53]

CONCLUSION

As Wally Broecker remarked back in 1998, "The climate system is an angry beast and we are poking it with sticks." Inherent in this metaphor is our lack of respect for nature and our ignorance or denial of the consequences of continuing to dump CO_2 and other wastes into the atmosphere at an accelerating pace. Feedback loops and tipping points make scientific certainty impossible: the best we can do is to put brackets around the range of possibilities for the fate of the planet—and our children's fate. The many numbers and statistics reviewed in these pages can be painful and even mind-numbing, but they will become more personal, and meaningful, when we see how they specifically impact our children. We'll turn to this in our next chapter.[54]

CHAPTER 2

Not Just Little Adults

INTRODUCTION

Since the late nineteenth century, our understanding of the biological vulnerability of the fetus and child has evolved dramatically, making it clearer than ever before that children have needs and concerns that warrant special attention when it comes to climate change.[1] This isn't simply sentimentality or just the idea that we have a moral obligation to care for the young, but the fact that young and developing bodies and brains bear the brunt of fossil fuel impacts. The World Health Organization (WHO) has estimated that more than 40% of the burden of environmentally related disease and more than 88% of the current burden of climate change is borne by children under 5 years of age, although that age group constitutes only 10% of the global population. That estimate for climate change was based on only three health outcomes (diarrheal disease, malaria, and malnutrition), so 88% is undoubtedly an underestimate.[2]

This chapter will take us on a focused journey through the many factors that contribute to the differential vulnerability of the young so we can understand why they are suffering the most from the burning of fossil fuel. It will end with an early case study in which the budding recognition of the vulnerabilities of children transformed society's view of the worth of children and powered an important policy change more than a century ago. This case study teaches us that the scientific understanding of today gives us a powerful weapon with which to tackle the most consequential challenge to children's health ever.

Laboratory and epidemiological studies have shown us that the special susceptibility of the fetus, infant, and child arises from a host of biological

Children's Health and the Peril of Climate Change. Frederica Perera, Oxford University Press. © Oxford University Press 2022. DOI: 10.1093/oso/9780197588161.003.0002

and behavioral factors all occurring at the same time. These factors include rapid growth, complex developmental programming, immaturity of detoxification and immune defense systems, limited ability to regulate body temperature during periods of severe heat, greater nutritional needs, and dependence on adult caretakers. Any one of these susceptibility factors would raise the risk from toxic exposures, but their combination greatly magnifies the harm to the developing fetus and child from air pollution, heat, malnutrition, viral infection, and stress linked to fossil fuel combustion and climate change.

Much of the research during past decades has focused on the brain and lung, which exemplify the beauty and complexity of development and the biological vulnerability to environmental and psychosocial stressors. Given the perfect storm of susceptibility factors, it is not surprising that a large body of recent animal and epidemiological research has shown that pre- and postnatal exposures to harmful stimuli damage the developing brain and lung and affect their functioning. We will see how these "toxic" exposures manage to elude the physiological defenses that nature has evolved to protect the fetus and child and the many ways they operate to cause harm. Then we will be better able to understand why the following chapter is titled "A Myriad of Health Effects."

THE BIOLOGY OF VULNERABILITY

The most important factor in the differential susceptibility of the young is the speed and elegant complexity of organ development in the fetus, infant, and child. The elaborately choreographed processes involved in early development are exquisitely vulnerable to disruption by "toxic" exposures of many kinds: adverse environmental conditions and physical toxicants, nutritional deprivation, and physical and psychological trauma and stress. [3]

Let's take the brain. By the time of birth, the basic structures of the brain are in place and most of the neurons, or nerve cells, that will be present in the adult brain have already been formed. One of the greatest marvels of biological engineering is the prenatal development of the brain—from the tip of the neural tube, then only about a tenth of an inch in size, to an organ with 100 billion neurons or nerve cells. This requires the brain to grow at an astonishing rate of about 250,000 nerve cells per minute, on average, throughout pregnancy.

A summary of the milestones reached before birth illustrates just how marvelous is this engineering feat. During the third week post conception, the neural tube forms. This tiny structure produces two types of cells that become the basis of the brain and spinal cord that together make up the central nervous system: the neurons which are the information processing cells in the brain and the glial cells that provide structural and metabolic support to the neurons. Neuron production begins 6 weeks post conception and is largely complete by mid-gestation. The neurons multiply at a dazzling speed, at a rate of over 4,000 cells per second. Regions of the brain that contain the cell bodies of neurons are gray in appearance, hence the name "gray matter."

Evoking the image of dancers bringing to life a complicated choreography, as the neurons are produced, they migrate to their appointed brain areas. To reach their destinations some cells must travel a distance as great as 1,000 times their own width. Once they have reached their target region, the young neurons must become part of information processing networks. They develop extensions (axons and dendrites) that allow them to communicate with other neurons by sending signals in the form of electrochemical waves: these cause the release of chemicals called neurotransmitters across the synapses or points of connection between two neurons. Axons are the principal means of sending signals from the neuron, while dendrites receive input from other neurons. Bundles of axons of neurons within one region of the brain form fiber tracts that connect with groups of neurons in other regions of the brain, forming the information processing networks. Beginning around the seventh month of gestation, the axons are wrapped in a fatty substance called myelin that, like insulation on a telephone wire, makes the transmission of electrochemical signals between regions efficient. Myelin is white in appearance; thus, fiber pathways of the brain are often referred to as "white matter." Eventually, billions of threadlike fibers crisscross the brain, forming complex networks that relay messages between different brain regions. These connections among nerve cells, referred to as the "connectome," make up the blueprint for the trillions of neural connections in the brain.[4]

From the ninth week of gestation through birth, the gross anatomy of the developing brain undergoes striking changes from a small, smooth structure to one with characteristic folds in the outermost layer or cortex. The elaborate folding of brain tissue makes it possible for large brains to squeeze into relatively small cranial vaults that will permit birth. By the end of the prenatal period major fiber pathways are complete. Communication between the cells that will make up the neocortex—the part of the brain that controls language, consciousness, and other higher functions—has begun.

Until recently, prenatal neurodevelopment in humans was a "black box." Most of the scientific knowledge about the steps in human brain development had come from studies of animal brains or post-mortem samples. That changed with the advent of new types of magnetic resonance imaging (MRI) that allowed heightened visualization of the connectome. Using this technology, in 2010, Christopher Smyser, a pediatric neurologist at Washington University in St. Louis, Missouri, found that babies born as early as 26 weeks already possessed immature forms of many of the functional brain networks seen in adults. Other researchers have successfully scanned the brains of fetuses over the course of pregnancy to follow the dynamic formation of the short- and long-range connections between different brain regions. So, although there is still much to learn about prenatal brain development and disturbance by toxic exposures, the black box has been cracked open.[5]

At the end of nine months, the brain is a functioning organ; however, it is still a work in progress. After birth, only a limited number of new neurons are produced; but cells that will become glia actively multiply, migrate to their preordained destinations, and mature. The processes of myelination of axons and the formation of synapses, launched before birth, continue through childhood. In a remarkable burst of growth, the number of synapses or connections that permit a neuron to pass an electrical or chemical signal to another neuron explodes in the first years of life—a phenomenon referred to as *synaptic exuberance*. In fact, the cerebral cortex produces most of its synaptic connections during the first several years after birth, at its peak creating 2 million new synapses every second.

By 2 years of age, a toddler's cerebral cortex contains well over a hundred trillion synapses. Exuberance peaks at 2–3 years of age. At that time, the level of connectivity throughout the developing brain far exceeds that of adults. Then comes *synaptic pruning*, when the brain refines its connectome to maximize performance by removing inefficient connections. At the peak of pruning, as many as 100,000 synapses may get cut every second. This refinement of the circuitry is shaped by activity and experience during early postnatal life and thereby forms the basis of learning and memory. In the case of little or no activity, some cells and connections are withdrawn, and those that are frequently used are strengthened. A scientist at Mt. Sinai in New York City, Robert Wright, explains that toxic exposures can interfere with the brain's ability to distinguish important connections from unimportant ones. For instance, environmental contaminants like lead can alter the normal trajectory of synaptic formation and pruning, derailing normal development of brain signaling networks. In addition to synapse formation and pruning, the most significant event in postnatal brain development is completion of the process of myelination so that mature brain cells are covered with this impermeable substance, forming the insulation that is essential for efficient electrical transmission between neurons. Neurodevelopment continues all the way through adolescence, during which there are significant changes: in myelinization, connections between synapses, and the distribution of chemical messengers that transmit signals from a nerve cell to its target.[6]

By what processes does this miracle of early brain development happen? Over the past three decades there has been tremendous progress in scientific understanding of this question. This has changed our fundamental models of how brains develop from a prescribed, linear model to one that is far more complex. Joan Stiles and Terry Jernigan at the University of California San Diego describe brain development as the product of a complex series of interactions between inherited, genetically intrinsic factors and environmental input. Neither by itself is determinative of whether or not there is a favorable outcome. These interactions between genes and environment occur throughout the course of brain development to promote the formation of new brain structures and functions. During the fetal period, genetic factors play a dominant role; but across the fetal period and extending into childhood, factors in the external world increasingly influence the course of brain development. Both genetically programmed gene expression and environmental input are essential for normal brain development, and disruption of either can fundamentally alter outcomes, as we shall see in a moment.[7]

Given the speed and complexity of the developmental dance it is remarkable that most young brains develop according to plan. But things can go wrong. Examples of neurodevelopmental disorders due to misadventures in early brain development include autism spectrum disorders (ASD) and attention deficit hyperactivity disorder (ADHD) which frequently co-occur. Imaging studies using MRIs of the brains of children diagnosed as having either or both conditions have found the same structural abnormality: lower volumes in a specific area of the cerebral cortex that is the center of learning, language, and memory. Other neuroimaging studies have attributed overlap in diagnoses of ADHD and ASD to shared abnormalities in white matter in the main nerve tract (the corpus callosum) that connects the right and left hemispheres of the brain. This thick bundle of nerve fibers, the first such tract to develop, enables communication between the two hemispheres and is a critical component of the connectome. The details of how these abnormalities occur and the various contributors are not well understood, but environmental as well as genetic factors clearly play a role.[8]

We have focused first on the brain, but the lung also illustrates remarkable speed of growth and complexity during the fetal and postnatal periods. While developing in the mother's womb, a fetus is dependent on mother's organs for oxygen that passes from the mother's blood through the placenta into the fetal bloodstream. At birth, the newborn requires a fully functional lung that can take in oxygen and remove carbon dioxide from the blood. To achieve this feat, the lung must develop from a tiny bud at about 30 days to an intricate tree-like system capable of delivering air to the gas exchanging units of the lung (the alveoli) at birth—another miracle of speed and precision.

Following development of the lung bud from the primitive gut, by 5 weeks two primary lung buds have been formed. These are the forerunners of the lobes of the lung. Between the 7th and 16th weeks, the lung buds undergo rapid branching to form the conducting airways. Between the 16th and 25th weeks, the alveoli have been formed and provided with a rich supply of blood vessels. This allows exchange of oxygen and carbon dioxide across these very thin, membrane-like cells.

Successful development and function of the lung requires not only the proper completion of the physical structure of the lung but also the biochemical development of the surfactant system, required for the stability of this very large surface area. The surfactants, composed of a mixture of fatty acids and proteins, are produced by cells lining the alveoli; their function is to provide a surface film that prevents collapse of the alveoli during exhalation. The surfactant system develops in the last trimester and reaches maturity by approximately 36 weeks.

Like the brain, the lung is not finished at birth. Lung growth continues after birth as the number of alveoli continues to increase to reach millions in number. The newborn has about 150 million alveoli; by age 8, that number is 300 million. The end result is an organ with a tremendously large surface area that is approximately 50–100 square meters in size, capable of exchanging oxygen and carbon dioxide across a very thin membrane.[9]

Having seen how the speed and complexity of early development place the fetus and child at great risk of harm from pollutants, let us look at the other factors that conspire to increase their vulnerability.

PREGNABLE BARRIERS

Many of us grew up thinking that the fetus lived a charmed life within the cocoon of the placenta, and we trusted that nature would have evolved a sound defense of the brain and other organs. But a big part of the scientific revolution has been the understanding of how easily many toxic exposures manage to elude the physiological defenses nature evolved to protect the fetus.

It turns out that the "impregnable" placenta and blood–brain barrier are myths—or, rather, that this idea may once have been true but is no longer. According to Philippe Grandjean, physician and epidemiologist at Harvard University, the placenta and the blood–brain barrier were probably quite sufficient during millions of years of evolution to safeguard the fetus, but new harmful chemicals have emerged that can manage to pass through the barriers.[10]

Most of us are not aware of what a complex, multitasking organ the placenta really is. The placenta not only provides uptake of nutrients and oxygen, but it also regulates body temperature, removes carbon dioxide and waste products via the mother's blood supply, fights internal infection, and produces hormones that support pregnancy. It is ingeniously designed for these purposes. The placenta is attached to the wall of the uterus and is connected to the embryo/fetus by the umbilical cord. It is composed of both maternal tissue and tissue derived from the embryo. From the first week of gestation the embryonic/fetal cells are separated from maternal tissues and blood by a layer of cells (trophoblasts) that transfer nutrients to the embryo and develop into a large part of the placenta. This interface not only serves as the means to extract nutrients from the mother's blood, but it also prevents contact between maternal blood and fetal tissue. The trophoblasts are organized into finger-like projections (villi) containing small blood vessels. The spaces between the villi are suffused with maternal blood so that nutrients, oxygen, and even certain antibodies can be transferred from the mother to the fetus, and waste products and carbon dioxide sent back from the fetus to the maternal blood. All of this happens in an organ that, at term, averages only 22 cm (9 inches) in length and about 2 cm (0.8 inch) in thickness.

When I was starting as a researcher into the environmental causes of disease, we were taught that the placenta was a nearly perfect barrier against toxic substances. That it would only allow the delivery of necessary nutrients and oxygen from the mother's bloodstream to the fetal circulation, while permitting the return of waste products back across the placenta to the mother's circulation for disposal. It was described as a very efficient system: oxygen and carbon dioxide were able to cross the placenta by passive diffusion; natural substances like glucose and amino acids for fetal protein synthesis were carried across by transporter proteins; and yet others like fatty acids and electrolytes, such as sodium and potassium, moved across by a combination of diffusion and transport. That much is still true—but the near-perfect barrier part was not.

As Grandjean describes it, when he was a medical student, he, too, was taught the basic physiology of the placenta that supported the view that adverse influences from the outside would not be able to bypass the filtering function represented by the placental barrier. But, for him, the beautiful hypothesis of the infallibility of the placenta was shattered by discoveries of the links between maternal alcohol abuse during pregnancy and fetal alcohol syndrome and between exposure of pregnant mothers to the German measles (rubella) virus and a number of congenital malformations. Both exposures targeted the developing brain as well as other organs. Since then, evidence of the permeability of the placenta to environmental chemicals and other toxicants has piled up, with the documentation by many researchers in the United States and other countries of their presence in cord blood, fetal tissue, or the placenta itself. Many of these invaders are byproducts of fossil fuel combustion.[11]

My own experience in the death of the "beautiful hypothesis" began when I saw the 1973 exhibition of photographs by Eugene Smith at the Institute of Contemporary Photography in New York City. The photojournalist had recorded the effects of exposure to methylmercury that had been released into waters around Minamata Bay, Japan, by a chemical factory. One of Smith's photos of a mother bathing her daughter moved me deeply. The metal had accumulated in fish consumed by pregnant women, often causing severe neurological effects in the fetus. Most mothers were not themselves affected. The photo (Figure 2.1), from another source, shows a woman holding a child who was a tragic victim of Minamata disease in 1973.

The second blow to the hypothesis was my shock in the early 1980s to learn that cord blood and placental tissue contained detectable levels of polycyclic aromatic hydrocarbons (PAHs) bound to their DNA—a sort of fingerprint of exposure. PAHs are present on the surface of fine particles emitted through the combustion of organic matter. Our subsequent studies confirmed the widespread presence of PAH-DNA *adducts* in newborn samples collected in the United States, Poland, and China. The average level of this fingerprint of prenatal exposure was highest in the cord blood samples in China, intermediate in Poland, and lowest in New York City, consistent with the relative levels of air pollution in the three countries. A subsequent survey of a nationally representative sample of pregnant women in the United States reported that a host of chemicals, including PAHs and metals such as mercury, were found in the blood or urine of virtually all the women tested.[12]

How did they get there? Now we know that, just like the natural molecules required for fetal growth and development, some environmental agents slip right through placental defenses by diffusion or active transport. These include bad actors that, like PAHs and mercury, are generated by fossil fuel burning

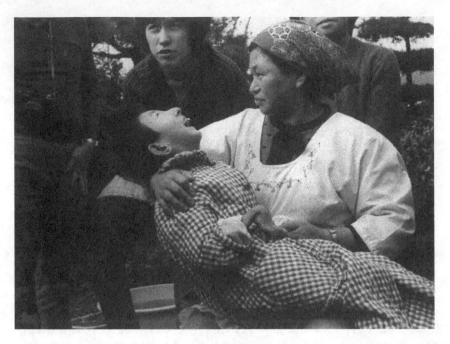

Figure 2.1 A woman holds a victim of "Minamata Disease," or mercury poisoning, in Minamata, Japan, in 1973.
Credit: AP/Shutterstock.

or are derived from fossil fuel. In general, the greater the lipid solubility of a chemical, the greater the placental transfer: PAHs, for example, are highly soluble in fats and other lipids and are thus able to make their way across the placental barrier with only limited filtering. Methyl mercury, the most toxic form of the metal, is actively transported. Maternal stress generates molecules that also cross the placenta: cortisol, the main stress hormone, is lipophilic (fat soluble) and easily crosses the placenta, as do cytokines, proteins that can trigger a harmful inflammatory response in the brain.[13]

The barrier that protects the developing brain likewise has been revealed not to be an ideal sentinel. Here also, my colleagues and I have direct experience of just how permeable the blood–brain barrier really is. But first, some background. The brain is the only organ known to have its own security system. Termed the "blood–brain barrier," this network of blood vessels allows the entry of oxygen and essential molecules while (theoretically at least) blocking harmful substances. Scientific research over the past century has identified and characterized the structure of the blood–brain barrier. It can be envisioned as a wall that protects the tissues of the brain. The imaginary bricks are the cells that line the interior of all blood vessels; but, in the blood vessels that form the blood–brain barrier, the cells are wedged extremely close together, forming so-called tight junctions. These tight gaps were designed to prevent foreign molecules and pathogens that may be present in the blood from slipping through, while allowing wastes and other substances to be expelled. Admission was to be allowed only to essential small molecules, fat-soluble molecules, and some gases so they could pass freely through the blood vessel wall and into brain tissue. Some larger molecules, such as glucose, were carried by transporter proteins that opened doors only for particular molecules. Thus, the blood–brain barrier evolved to operate as the brain's sentinel and "bouncer."[14]

However, research has shown that this protection is far from complete, with many new harmful chemicals quite easily breaching the barrier. The same characteristics that allowed PAHs, mercury, and stress-related molecules to evade the placental defense also permit them to reach the developing brain. As we will see in the next chapter, the consequences can be serious. All the more so because these toxic invaders show great versatility in derailing normal neurodevelopmental programs.

IMMATURE BIOLOGIC DEFENSE SYSTEMS

The fetus, infant, and child are susceptible to air pollution, mercury, lead, and other fossil fuel contaminants as well as to climate change impacts (heat, infectious agents, and stress) because of the immaturity of their biologic defense systems; thus, they may be affected by exposures that have no apparent effects in adults. These biologic defenses include enzyme systems that metabolize toxic chemicals like PAHs into more benign forms that can be safely excreted from the body. Another set of enzymes repairs DNA damage caused by environmental pollutants or a reactive form of oxygen generated by maternal stress. This is critical because, unless the damaged segments are repaired promptly and accurately, the DNA replication machinery will misread the code, potentially resulting in mutations, cancer, and other diseases. When DNA damage from chemicals is detected, the DNA repair team is called in: these enzymes cut out the corrupted segment of DNA and replace it with the correct sequence. Like the detoxification enzyme system, DNA repair enzymes generally work well in adults, but their efficiency is limited in fetuses and children. My own research has demonstrated heightened susceptibility of the fetus to DNA damage in white blood cells in cord blood. Comparison of the levels of PAH–DNA adducts in paired maternal and umbilical cord white blood cells showed that, despite the estimated 10-fold lower PAH exposure of the fetus compared with the mother, the levels of PAH–DNA adducts were similar.[15]

Immune defenses are also limited in the young. The fetus and newborn are highly susceptible to infection. Although certain types of immunity in the form of ready-made antibodies are passed through the placenta from the mother, they only protect against some diseases but not others. The fetal immune system itself develops along two lines. The first line of defense is *innate immunity*, composed of white blood cells that act like security guards patrolling the circulatory system to identify, engulf, and destroy invading bacteria and viruses. The second line of defense is *acquired immunity*, in which specialized white blood cells create an immunological memory after an initial response to a specific infectious agent; upon re-exposure the cells recognize the invader and mount a defensive response. The innate immune system is muted at birth, which makes the newborn—and particularly the premature baby—susceptible to bacterial and viral infections. The second line of defense is likewise immature in the fetus and newborn: their antibody response is at a lower intensity compared to older children and adults. The innate and adaptive immune systems gradually mature during infancy and early childhood so they become better armed against infectious agents. Over time, protection provided by the immune response increases so that young adults suffer fewer infections than at earlier stages.[16]

As discussed in the next chapter, a tragic example of an infection eluding the fetal immune defenses is the Zika virus, which evades fetal innate immune responses to move into the fetal brain. Malaria, dengue fever, and tick-borne diseases such as Lyme disease also take advantage of immature immune systems to disproportionately affect children.[17]

VULNERABILITY TO HEAT

Toddlers and other young children are at particular risk of heat-related illness (hyperthermia). I have seen a child experiencing the early stage of hyperthermia or *heat stress*; she was tired, breathing rapidly, and had a rapid pulse. Fortunately, she was quickly treated with fluids, cooling compresses, and rest in a cool place. In the more severe stage (*heat stroke*), a child may vomit, feel dizzy, and even have a seizure. If untreated, heat stroke can be fatal. More heat-related deaths among infants are reported during heatwaves. During persistent episodes of hot weather children are also more likely to suffer from respiratory disease, kidney disfunction, fever, and an imbalance of electrolytes such as sodium, potassium, and calcium that are vital in maintaining optimal body temperature and fluid balance.

There are several reasons for the heightened vulnerability of the young to severe heat—both biological and behavioral. Infants and children have less ability than adults to regulate body temperature in conditions of severe heat. Children spend more time outdoors and participate in more vigorous activities than adults, which results in greater exposure to outdoor heat. They are generally unable to recognize or signal the early signs of distress and must rely on adults for hydration and care. For these reasons, along with older adults, children ages 0 to 4 years are at particularly high risk of heat-related illness and death.[18]

GREATER NUTRITIONAL NEEDS AND OTHER VULNERABILITIES

Climate-related drought and crop failure particularly affect the young by depriving them of the macronutrients (protein, fat, and carbohydrates) that are the main sources of calories ("energy units") and of the micronutrients (minerals and vitamins) required for healthy growth and organ development. Climate disasters that affect nutrition can therefore have devastating effects on infants and children. Newborn infants have low stores of fat and protein and can only cope with starvation for short periods of time. Throughout infancy and childhood, nutritional requirements are higher per kilogram of body weight than at other developmental stages. Because of their particularly rapid growth, infants have the greatest nutritional demand: a 1-month-old infant requires about four times the number of calories per kilogram of body weight compared to an average adult. Energy needs remain high throughout the early years: a child 1–3 years of age requires more than three times the number of calories per kilogram of body weight than an average adult.[19]

Water is also considered an essential nutrient. Infants and children require two to three times the amount of fluid than adults. This is because their larger surface area in relation to weight allows greater water loss through the skin, and they lose more water through the lungs because of their higher respiratory and metabolic rates.[20]

The developing brain is especially vulnerable to nutritional deficits during pregnancy, infancy, and early childhood. From conception to 3 years of age, the rapidly growing brain requires more nutrients than at other stages of development. In addition to the B vitamin, folate, which is particularly important for brain development prenatally, key nutrients for brain development during this period include protein, certain fats, iron, zinc, copper, iodine, selenium, choline, and vitamin A. Although malnutrition is linked in most minds to stunting of the body and photos of pitiably malnourished children, as we will see in the next chapter, stunting of the brain is the more serious outcome, affecting health and functioning over the child's life.

Children are biologically more vulnerable to physical injury and psychological trauma from displacement due to weather disasters. They are also dependent on adult caregivers who themselves may be affected by injury or psychological trauma from forced migration.

A further vulnerability comes from the long remaining lifetimes of children, during which the biological insults that occurred during fetal and early child development become newly manifest as heart disease, lung disease, or conditions of older age such as Parkinson's or Alzheimer's disease. Many diseases have a long latency period, sometimes requiring decades to develop. In these cases, we think of harmful exposures during fetal and early childhood development as "seeding" adult disease. In other cases, early disease or neurodevelopmental problems may persevere, affecting health and functioning throughout the life course. Examples include certain types of asthma and mental health problems in adolescence that may persist into adulthood. Another is IQ loss due to malnutrition during the first 1,000 days of life, which affects the ability of children to learn and as adults to earn. Epidemiological studies have established that IQ is correlated with lifetime earnings, so there is an economic cost as well that may impact future generations.

THE HOW OF TOXIC INTERFERENCE

As we have seen, infants and children have a dynamic physiology that is turned up to "high", so their needs for energy, water, and oxygen are higher than at older ages. As a result, they have far greater exposure per unit of body weight to toxic pollutants that may be present in their food and water. Toxicants that are carried in food are delivered at two to four times higher rates in children than in adults, and those in water are delivered at two to five times the adult rate. Toxic pollutants that are emitted into the air from fuel combustion are often deposited on land and water, accumulating in crops and fish consumed by children. In addition to their toxicity to the developing brain and their interference with the function of natural hormones, toxicants like PAHs, lead, and mercury are harmful to the immune system, so the infant or child will have even less ability to mount a defence against viruses and other infectious agents.[21]

Another biologic difference is the higher breathing rate of infants and young children, resulting in a greater intake of any pollutants present in the air. Children also tend to be more physically active than adults, further increasing their intake. Therefore, environmental toxicants found in the air, such as air pollutants, molds and pollen, are directly delivered to children at higher doses than to adults. Their disproportionate exposure, plus the increased concentrations of these toxicants due to climate change, are clearly factors in the high incidence of childhood asthma and hospitalizations due to asthma.[22]

Toxic exposures show considerable versatility and inventiveness in disrupting normal developmental processes in ways that lead to disease and impairment. Knowledge of the mechanisms involved is limited at this time but laboratory and epidemiological studies using biomarkers have suggested how environmental toxicants and stress can derail development. A look at several of the known or likely mechanisms reveals the ingenuity of toxic invaders. Moreover, a chemical toxicant or stressor may act through several of these mechanisms—another aspect of their versatility.

We have already seen how air pollutants and stress can cause prenatal DNA damage that may lead to an alteration in the DNA sequence (a gene mutation): the result is disruption of the normal process by which the DNA code is translated to proteins which are the essential building blocks of the body. The resultant proteins may be nonfunctional or abnormal, with potentially serious consequences for health and brain development. However, there is another, ingenious way that toxic exposures disrupt protein production; this occurs by turning certain genes on or off at the wrong moment in early development but without altering the sequence of the DNA making up the gene. This type of damage is termed "epigenetic" (derived from the Greek for "above the genome"). DNA methylation is the most extensively investigated of the epigenetic mechanisms. It is generally thought that the addition of a methyl group at a specific site on DNA effectively silences the gene; conversely, removal of a methyl group activates expression of the gene. The metaphor of a light switch is often used in this context. The end result is the same as from a gene mutation and may have grave consequences for the developing fetus and child.

The epigenome is susceptible to dysregulation throughout life. However, it is thought to be most vulnerable to environmental factors in the embryonic stage, which is a period of rapid DNA synthesis and epigenetic remodeling. In an intricate choreography, after fertilization of the egg, DNA methylation patterns in a large set of genes are erased and then reestablished before birth. But the epigenome is not static after birth because cell methylation patterns in the non-reproductive cells adjust in response to developmental and environmental factors. Numerous studies, including in newborns, have shown that air pollutants such as PAHs, particulate matter, ozone, and mercury, and stress are among the environmental exposures capable of altering normal epigenetic programming, with possible consequences for the health of the child. There may even be impacts of early-life exposures to air pollutants, nutritional deprivation, and stress on future generations via the transmission of epigenetic changes.[23]

Another mechanism involves oxidative damage, which occurs when reactive species of oxygen produced by toxic exposures such as air pollution, mercury, and stress overwhelm the body's detoxification and repair systems, damaging DNA, protein, and other components of the cell. Oxidative damage can result in gene mutation or in altered gene expression through DNA methylation changes, as we have seen with chemicals like PAHs. A third mechanism is interference with the immune system to produce inflammation. Chronic stress during early development is linked to immune system malfunction, resulting in an excess production of inflammatory proteins, with potentially serious lifelong consequences in terms of chronic disease. The same inflammatory response can be triggered by exposure to air pollution, and the two exposures have been shown to interact, thus increasing risk of developing asthma and other conditions. Mercury also stimulates the immune system to release inflammatory proteins. Displaying even greater versatility, toxicants can inflict damage by interfering with chemical messengers, or neurotransmitters, that allow neurons to communicate with each other and with muscles and glands. Other toxic exposures can interfere with natural hormones such as brain derived neurotrophic factor (BDNF) and sex hormones that are intricately involved in early brain development. In fact, studies in humans, including newborns, have found that certain industrial chemicals and air pollutants are able to target these critical proteins and affect their levels and functioning in the fetus.[24]

One has to marvel, ruefully, at the versatility by which environmental exposures traceable to pollution and climate change are able to inflict harm during the exquisitely sensitive early stages of development—particularly since the same exposure can act via several of the mechanisms involved. Given the inability of science—so far—to develop ways to block these mechanisms, it is clear that preventing the exposures from occurring in the first place is the only solution.

HOW EVOLVING SCIENCE LED TO A POLICY CHANGE TO PROTECT CHILDREN

The rapidly evolving science concerning early development has led to a corresponding evolution in the perceived value of children and society's responsibility to them, and there are lessons from the recent past that show how an engaged—and enraged—citizenry can make sweeping changes in the name of protecting children's health.[25]

In the late nineteenth century in the United States and Europe, the view of children began to evolve from "economically valuable" to "emotionally priceless" and therefore worthy of full protection from harmful exposures. Whereas in the pre-modern or colonial period the young had been viewed by all socioeconomic classes as adults in training and as an integral part of the family's economic well-being, by the late nineteenth century middle- and upper-class parents had adopted the view of children as innocent and vulnerable, requiring nurturing and protection from harsh adult realities. Portraits of American children vividly demonstrate the change in how children of the middle- and upper-classes were viewed between the seventeenth century, when children were regarded as "little adults," and the latter part of the nineteenth century, when children had become "priceless." We see the Mason children painted in 1670 (Figure 2.2a), stiffly enveloped in miniature versions of adult clothing. We see another child, Kate Rosalee Dodge (Figure 2.2b), painted 200 years later, well-dressed and holding a book, clearly experiencing a cosseted and protected childhood.

The romantic ideal of childhood, embodied in the portrait of the Dodge child, was a largely upper-class concept. Poor parents in the latter 1880s had a different view: that their children had a reciprocal obligation to contribute to their household's well-being. By the late 1800s, however, the construction of the priceless but economically worthless child had been accomplished among the American urban middle class. By the 1930s, lower-class children joined their middle-class counterparts in a world where the sanctity and emotional value of a child made child labor taboo. Today, almost everywhere, children are valued as precious. This value shift brought with it the moral responsibility of protecting the health and future well-being of children, a responsibility that is now recognized by virtually all cultures and societies. It is worth noting, however, that largely due to perceived or real necessity there are still more than 200 million working children in the world, most in the developing countries.

The passage of the New York Child Labor or "Factory Act" of 1886 was a major milestone in the evolution of the "priceless child" in the United States and the first concerted use of the then available science on developmental vulnerability to drive a major policy change concerning children's health. We can draw from it several lessons for the present issue of climate change and children's well-being, to which we shall return later in the book. One of those lessons is the power of data on the special vulnerability of

Figure 2.2 (a) The Mason Children, 1670. (b) Kate Rosalie Dodge. 1854.

Credit: (a) Anonymous, oil painting, de Young Museum Fine Arts Museum of San Francisco; (b) John Wood Dodge, Watercolor on ivory, American, Metropolitan Museum of Art, Morris K. Jesup Fund, 1988.

children to awaken the dormant sense of responsibility in adults, especially policymakers.

In their landmark campaign to secure greater protection of children who were employed in factories, the reformers wrote, often movingly, about the biological and psychological vulnerability of children. Their effective communication in newspapers, periodicals, and reports accelerated understanding in the nineteenth century of the uniquely vulnerable child and the recognition that children were of inherent value and deserved the care and protection of adults.

The campaign unfolded in the social context of the second half of the nineteenth century in the United States, when the fabric of society was changing rapidly due to industrialization and immigration. By 1870, the population of New York City had swelled to over a million people, of whom half were foreign-born and mostly poor. The mortality rate for children under 5 was a staggering 52%. As now, there was a sharp economic and educational discrepancy between rich and poor in New York. The wealthy mercantile class and the older, established families enjoyed prosperity, educational advantage, and positions of power—in sharp contrast to the families of the poor who were largely uneducated and foreign-born.

Demand for child labor in New York City was high, particularly in textiles, canning, mining, and street peddling, where children as young as 6 or 7 worked long shifts. Nearly 100,000 children were employed in factories and shops in New York City and its suburbs. Photos by Jacob Riis showed poor working children of New York City whose lives contrasted dramatically with the bucolic lives of children of the previous century, many of whom worked on family farms or as apprentices in training for a profession. These photographs of poor urban children aroused the passion of reformers to protect children from harm and exploitation in the workplace (Figure 2.3a,b).

In the latter part of the nineteenth century, most of the evidence on the harm to children from laboring in dangerous and polluted workplaces did not come from well-designed epidemiological studies (studies of the patterns and determinants of health and disease in defined populations) or molecular epidemiological studies (studies that incorporate biologic markers of exposure, preclinical response, and susceptibility), but from direct and indirect observation or the experience of physicians, toxicologists (who study how chemicals interfere with the normal function of a biological system), factory inspectors, sociologists, and psychologists. Their reports describing working conditions and illness in children soon began to appear in the public press.

Figure 2.3 (a) Children working as knitters in a textile mill in Tennessee. (b) Poor children on Coney Island, New York.
Credit: (a) Photos by reporter and reformer, Jacob Riis. CBW/Alamy Stock Photo; (b) Jacob A. Riis (1892) Museum City of New York.

In 1873, an article in *Harpers New Monthly Magazine* described the work environments of tobacco manufactories in which 10,000 children were employed: "by far the most noxious environment in which the under-ground life in these damp caverns tends to keep the little workers stunted in body and mind." Claire DeGraffenreid, an investigator with the US Bureau of Labor, she detailed the many adverse effects of tobacco poison: "extreme nervousness, maladies like St. Vitus Dance, physical weakness, disordered digestion, heart action impaired, strength sapped; the mind is excited, often the passions are inflamed and the moral sense deadened." An assistant factory inspector in Illinois, Alzina Parsons Stevens, observed: "To know how a child is affected who breathes this atmosphere all day, bent over a tobacco bench, take up her hand and examine the shrunken, yellow fingertips, the leaden nails; lift her eyelid, and see the inflammation there; examine the glands of her neck, her skin; lay your hand upon her heart and note its murmur. Nor does the injury to the girl child in the cigar factory end with herself. The records of the medical profession show that women who have worked in the tobacco trade as children are generally sterile. When their children are not stillborn, they are almost invariably puny, anemic, of tuberculous tendency, the ready prey of disease."[26]

In short, wrote de Graffenreid, having surveyed all the various occupations in which children were working: "[M]any children are engaged at tasks too great for their physical strength, becoming consumptive in consequence or suffering serious bodily harm. . . . These years when mind and body are susceptible of the healthiest growth are spent in a monotonous round of indoor drudgery which undermines the constitution, stunts the intellect, and debases the higher nature."[27]

Many writers underscored another threat as insidious as the physical toxicants: the psychological harm and moral degeneration resulting from exhausting labor, scanty food, and lack of education. They saw poor uneducated children, stunted in body and mind, as both a tragedy and a potential threat to society.

The reformer-writers, mostly women, harnessed moral principles to science in order to shift public opinion away from the acceptance of child labor. They argued that the practice of child labor was an evil not tolerable in a civilized society. They used every weapon of biblical rhetoric, irony, and fear, not only to arouse pity for these unhappy children, but also to conjure a future of an ignorant and illiterate citizenry, physically and morally degenerate as a result of the toxic and corrupting environment in which they were forced to spend half of their "unnatural lives." In periodicals like *The Arena*, reformers characterized child labor as child slavery, with indignant references to the Bible: "To force a child whose only inheritance is a weak constitution into employments which require the fullest development of mind and body is an act which out-Herods Herod."[28]

Nor were they afraid to tweak the conscience of the privileged class, as, in the words of de Graffenreid, "Think of it, parents, who kiss your pampered darlings of nine and ten years in rosy slumber tucked away at eight o'clock in the soft, warm bed after a day of romp, wholesome food and wisely managed study! On Sunday mornings the writer has seen at their homes scores of cash-girls and boys heavy-eyed, listless, dragging their tired limbs or asleep in the stupor of exhaustion. Where are the graces, the joys, the innocence of childhood?"[29]

These accounts engaged the public—particularly privileged, high-status people with political power. They were concerned about the plight of poor children; but they also feared that the surge in the ranks of poor, uneducated, and unhealthy child workers posed a threat to themselves. They therefore had a dual view of the problem of child labor: it was both a wrong that needed righting and a threat to their own families' well-being and future security. (This duality previews the common attitude toward climate change today, as people whose children have not been directly affected—at least visibly—may feel some passive sympathy for "those other children" who have been harmed but are truly activated by their desire to provide a viable and secure present and future for their own children.)

In the end, the reformers were successful in using these accounts and levers (both altruistic and selfish) to power a new social and political movement. In this, they were assisted by a cooperative press that facilitated the rapid diffusion of scientific knowledge to the public and among reformers in different states. The word went out in articles published in newspapers, periodicals, and widely circulated reports.

Although the abolishment of all child labor in the United States did not occur until 1938, by the late nineteenth century science had clearly emerged as a player in policymaking. The nineteenth-century reformers were successful because they had at hand some empirical knowledge of the immediate, life course, and even transgenerational impacts of toxic chemicals and psychosocial stressors. They were able to communicate that knowledge in compelling ways that engaged public support in reform. This early success in policy reform through raising awareness of children's vulnerability—even using the limited data available at that time—encourages us today as we confront the greatest threat ever to children's health and wellbeing.

CONCLUSION

Scientific research—especially in the past decades—has resulted in a quantum leap in knowledge on the vulnerability of early development, with a corresponding paradigm shift in society's view of the value of a child. Before we can discuss how this robust evidence is fueling new efforts with our current challenges, we must first shift our focus from *why* children are so susceptible to just *how* they are being affected and to whom we should pay the closest attention.

CHAPTER 3

A Myriad of Health Impacts from Fossil Fuel

"Myriad"—in classical history a unit of 10,000—is technically an over-statement but captures the notion of a great number of health effects attributable to fossil fuel combustion. Building on the previous chapter, here we will see the many health and developmental impacts now being inflicted on the young, beginning before they are born, by the toxic air pollutants and climate-altering gases released by the production and combustion of fossil fuel. By looking at these threats holistically—effectively moving climate change and derived air pollution out of their traditional silos—we will come to a full reckoning of the harm to children that results from a carbon-based economy and we will understand the deep urgency of the matter. This will be the basis for thinking together about how to advance the needed comprehensive and equitable policies to protect this vulnerable group.[1]

CONSIDERING AIR POLLUTION AND CLIMATE CHANGE TOGETHER

It makes sense to jointly consider these twin threats: they share a common source and their co-occurrence magnifies the harm to children's health. Moreover, understanding the health effects of air pollution has provided critical support for action on climate change. Because the health effects of climate change have been less easily quantified, in 2009 the benefits of reducing CO_2 and other greenhouse gas emissions began to be framed

Children's Health and the Peril of Climate Change. Frederica Perera, Oxford University Press. © Oxford University Press 2022. DOI: 10.1093/oso/9780197588161.003.0003

in terms of the health "co-benefits" from reduction of air pollution. This approach made sense in an era when powerful interests were furiously seeking to cast doubt on climate change science itself. Moreover, as a 2009 report in the journal *Lancet* noted, "The co-benefits to health arising from action on climate change are not widely appreciated. A greater awareness might sweeten the otherwise bitter taste of some climate change policies." There were other advantages as well: unlike CO_2, which is an invisible, odorless gas, much of the public had seen, smelled, and experienced air pollution and knew at least something of its health impacts. A solid database had been built over many decades on the direct health effects of air pollution. The effects could also be localized, whereas changes in CO_2 emissions locally could have effects anywhere in the world and the consequences were not distributed predictably. Flashing forward to the present, air pollution is now seen as a public health emergency in its own right.[2]

There has been a large gap, however, in the accounting of the benefits of reducing air pollution. Since 2009, the health outcomes that have received the most attention have been the avoided deaths and illness in adults from improved air quality. It has long been clear to me that we needed to delve much deeper into the health benefits. This meant a new focus on the improvement in children's health and well-being. It also meant considering the harmful interactions between air pollution and climate change and presenting the benefits holistically.

The next pages describe the current state of knowledge on air pollution and climate change, first separately and then as they combine to impair children's health, often with long-term consequences. The science presented will clearly show that children are being multiply affected right now from concurrent exposure to air pollution, heat, higher levels of plant allergens, smoke from fires, severe storms, drought, and other impacts of climate change. At chapter's end, we will be able to picture a young child who developed asthma as a result of high exposure to air pollution and then suffers asthma attacks triggered by plant allergens or smoke from forest fires. We will see a baby born preterm due to combined exposure to air pollution and heat during pregnancy who then must struggle to succeed in school and is prone to infections, asthma in childhood, and long-term intellectual disabilities. Or an adolescent suffering from mental health problems following a weather disaster, compounded by the effects of neurotoxic air pollutants. To do this, we must delve into the science accumulated by hundreds of dedicated investigators over the past several decades.[3]

Here are two sobering statistics: today, approximately 2 billion children in the world breathe toxic air at levels exceeding guidelines set by the World Health Organization (WHO), most of which is generated by the burning of fossil fuel. More than half a million children die before their fifth birthday every year from air pollution–related causes, and an even greater number are afflicted by lasting damage to their developing brains and lungs.[4]

Before we plunge into specifics on the health effects, there are several caveats to keep in mind.

1. Almost all diseases have multiple causes: they result from genetic, environmental, and social factors, often playing off one another. No single factor fully explains a disease. The rationale for focusing on environmental exposures is that, unlike genes, once they are identified as harmful these exposures can be prevented. Furthermore, the production of disease almost always requires multiple factors acting in concert: each is necessary but not sufficient to cause disease. Therefore, if one significant contributing factor is removed, disease may not occur.

2. Epidemiology by itself cannot establish causality with respect to air pollution and disease. That's because causality cannot be proved in observational studies of populations; it can only be inferred. Proving causality would take studies that place groups of pregnant women or children in aerosol chambers for months or years to breathe differing levels of pollutants—clearly an infeasible, not to mention unethical, approach. Laboratory experiments are therefore undertaken to document the effect of a known amount of the pollutant and shed light on the mechanisms by which the pollutant is acting. The US Environmental Protection Agency (EPA) and most other regulatory agencies around the world are required to act to restrict an environmental exposure to "safe" levels if it is "known or *likely to be* causally related" to an adverse health outcome, considering all the epidemiological and experimental evidence. Of course, we often see that politics can intervene to steer the decision toward or away from action.

3. Air pollution, like other significant risk factors for poor health, such as smoking and obesity, is rarely if ever indicated in official records as a cause of death, hospital admission for heart attack, or emergency department visit for asthma. Therefore, in observational studies, epidemiologists determine the relationship between air pollution (say, fine particulate matter or $PM_{2.5}$) and the incidence of the illness or death in a study population and calculate the increase in risk per unit of

exposure: this is the *concentration-response function*. Models are run to estimate the number of deaths or disease linked to the pollutant relative to the background that would occur in the absence of human-produced emissions. This is the estimated number of excess deaths or disease, referred to as the *public health burden*. The key word is "estimated," given the various assumptions in the model. Because this exercise rests on observational research rather than experiments, the word "association" is used to describe the relationship between air pollution exposure and a health effect.

4. As we will see shortly, for some health outcomes there have been only a few studies of adequate size and design to permit causal inferences. In addition, results across studies are often inconsistent. This is understandable because of differences in levels of exposure, the racial/ethnic and socioeconomic makeup of the population studied, the methods used to assess outcomes, or the failure in some studies to account for *confounding* by an unmeasured factor that explains the apparent association with the exposure of interest. Where a sufficient number of well-conducted studies are available, statistical analyses (such as meta-analyses) are performed to combine their results.

Readers will notice the often sizeable estimated monetary costs of childhood illness attributable to air pollution. But there are several caveats here as well: impressively steep as these costs seem, they are almost always underestimates, as we will see in Chapter 7, "Solutions Now." More importantly, the very exercise of calculating avoided economic costs may seem to imply that the price tag is the most important consideration. We all know that the burden borne by parents and children facing an illness or disability includes incalculable physical and psychological suffering that dwarfs the monetary cost. We perform these cost exercises, though, because decisionmakers generally require them to buttress their claims for (or against) action. The key is to make these as holistic and comprehensive as possible.

Mortality

Most reports on the human health burden of air pollution and the benefits of corrective policies have focused on excess mortality and cardiovascular disease in adults. That has left out many health effects in children often initiated in utero, including infant mortality, preterm birth, low birth weight, impairment of cognitive ability, behavioral problems such as attention deficit hyperactivity disorder (ADHD), autism, asthma attacks, and new cases of childhood asthma. These are all too common. The focus on mortality is certainly understandable, however. According to the Lancet Commission on Pollution and Health of 2017, of which I was a member, air pollution remains "one of the great killers of our age." Since that report sounded its loud alarm, new research has estimated that more than 10 million people died prematurely in a recent year from fossil fuel pollution, and exposure to particulate matter from fossil fuel emissions accounted for 18% of global deaths—almost 1 out of 5. Regions with the highest concentrations of fossil fuel-related $PM_{2.5}$—China, India, parts of the eastern United States, Europe, and South-East Asia—had the highest rates of mortality. The annual toll in premature deaths due to air pollution is more than a million in Europe and over 480,000 in North America. The global loss of life expectancy caused by air pollution due to fossil fuel use is higher than by smoking, infectious diseases, or violence and has been likened to a pandemic.[5]

Almost all the deaths are in older adults, but air pollution contributes to almost 500,000 deaths among infants in their first month of life, largely from complications of preterm birth, pneumonia, and other respiratory infections. Newborns who are already susceptible to infection because of their immature immune systems are vulnerable to a "second hit" from air pollution. By weakening the lung's immune response to respiratory infection and then causing further inflammation, air pollution causes a more serious course of illness. Because of the increasing trend in outdoor air pollution levels globally, it is estimated that under-5 mortality could be 50% higher than—or even double—the current number by 2050. The grim prediction is that, at the present rate of emissions, by 2050, outdoor air pollution will become the leading cause of child death. The good news is that, in the world's largest cities, as many as 153 million premature deaths linked to air pollution could be avoided by century's end if governments act promptly to reduce fossil fuel emissions and limit global temperature rise to 1.5°C (2.7°F) of warming. Said the lead author on this study, "Hopefully, this information will help policymakers and the public grasp the benefits of accelerating carbon reductions in the near term, in a way that really hits home."[6]

Birth Outcomes

Let us turn now to the chronic effects of air pollution on child health starting at birth. Unlike mortality, these are rarely in the headlines and are less recognized by the public as a consequence of air pollution exposure. Globally, each year a staggering 15 million babies are born preterm, defined as before 37 completed weeks of gestation, and this number is rising. Even in the United States, of the almost 4 million children born every year, more than 380,000 (about 1 in 10 babies) are preterm. No single social or environmental factor is uniquely responsible; however considerable research has shown air pollution to be a significant contributor. Globally, 2 million premature births in 2019 were attributed to ambient exposure to $PM_{2.5}$, mostly, but not all, in developing countries. In the United States, almost 12,000 preterm births were attributable to ambient $PM_{2.5}$ in the same year. Preterm birth complications are the leading cause of death among children under 5 years of age, with more than 1 million babies who are born preterm dying shortly after birth, while many more preterm babies suffer some type of costly lifelong disability.[7]

Another serious birth outcome linked to air pollution is low birth weight (<2,500 grams or 5.5 pounds). About 20 million low-birth-weight babies are born worldwide every year. While almost all are in developing countries, in the United States more than 310,000 babies are born low-birth-weight every year. An exhaustive review of 48 studies comprising 32 million births in the United States concluded that there was overwhelming evidence that exposure to $PM_{2.5}$ is associated with increased risk of both preterm birth and low birth weight.[8]

Exposure to air pollution and higher temperatures may combine and even interact to increase risk of preterm birth. In an analysis of more than 215,000 live births in Guangzhou, China, exposure to $PM_{2.5}$ and even moderate heatwaves interacted synergistically, such that their combined effect on the risk of preterm birth was greater than additive.[9]

Disorders due to preterm birth and low birth weight are not only the leading causes of infant mortality, but also contribute to lifelong consequences of being born too soon or too small. Preterm babies are at higher risk for lower respiratory infections, other infectious diseases, asthma in childhood, and long-term intellectual disabilities, IQ loss, ADHD, autism, anxiety, and depression—often extending into adulthood. The likely developmental sequelae for low-birth-weight infants include mild problems in cognition, attention, and movement.[10]

These health effects are costly to individuals, families, and society. Economic data from other countries are limited, but we know that, in the United States, the cost per case of preterm birth ranges from $70,000 (counting only short-term costs) to $325,000 (including the loss of lifetime earnings due to even modest IQ reduction). In the United States, the estimated short-term cost per case of low birth weight ranges from $15,000 in full-term babies to greater than $110,000 for very low-birth-weight babies. The true costs are far higher.[11]

Disorders in Brain Development

Given the vulnerability of the developing fetal, infant, and child brain, it is no surprise that research during the past two decades has produced convincing evidence that early-life exposure to combustion-related air pollutants (and many chemicals derived from fossil fuel) adversely affect children's cognitive and behavioral development starting before birth.

Neurodevelopmental disorders (conditions in which the development of the brain is disturbed, affecting a child's behavior, memory, or ability to learn) are seen in a large and growing number of children globally. Most of us know a child with some form of ADHD—a condition that affects 1 of 10 children—or autism spectrum disorder, which affects 1 in 44 children in the United States. The prevalence of autism in Asia, Europe, and North America as a whole is similar—between 1% and 2%. Globally, an estimated 14% of children and adolescents suffer from a mental disorder. Rates of all these conditions have escalated in recent decades. These increases have not been explained by better diagnosis alone, turning the spotlight onto environmental exposures.[12]

Cognition

The term "cognition" is very broad, encompassing many aspects of intellectual functioning and processes such as perception, memory, judgment, and reasoning. A number of cohort studies in the United States, Europe, and Asia have evaluated the associations between early-life exposure to $PM_{2.5}$ or traffic-related air pollutants (black carbon and nitrogen dioxide) and cognitive abilities in infancy and childhood. These studies have largely relied on estimates of exposure based on air monitoring data, type of land use, distance of the residence to roadways, or traffic density. The researchers have accounted for factors like income, age, race/ethnicity, and education. Several of the studies linked $PM_{2.5}$ exposure to reduced IQ; however, the results for this pollutant have been mixed. In contrast, most of the studies on traffic-related pollution have reported associations with decreased mental development in young children, including reductions in children's memory and other measures of IQ. Where carefully assessed, sex-specific effects of air pollution on cognition have been apparent, with boys more affected.[13]

My colleagues and I have studied polycyclic aromatic hydrocarbons (PAHs) for many years with respect to cognitive outcomes in populations living in New York City (NYC); Krakow, Poland; and Chongqing, China.

Our research has found consistent associations between prenatal PAH exposure and cognitive problems across these different populations. In our main NYC cohort comprised of low-income African American and Latina mothers and children, prenatal exposure to PAH was measured in personal air using small backpack monitors or by PAH-DNA adducts in cord or maternal blood collected at the time of delivery. As the children moved through infancy and childhood, research staff repeatedly assessed their cognitive development using age-appropriate tests. PAH exposure was associated with delayed mental development at age 3 and reduced IQ at ages 5 and 7. Using the same research approach in our Polish cohort, we found a similar effect of prenatal exposure to PAH on children's intelligence scores. The reductions in IQ and intelligence scores seen in our studies in NYC and Poland with high prenatal PAH exposure were comparable to the effects of low-level lead exposure. Our Chinese study also found decrements in IQ in children who had combined prenatal exposure to PAH and secondhand smoke.[14]

The Chinese study, conducted in Tongliang District in Chongqing Municipality, allowed us to take advantage of a "natural experiment" and directly assess the neurodevelopmental benefits of *reducing* air pollution levels. The main exposure source was a centrally located coal-fired power plant entirely lacking in emission controls. The government had announced its decision to shut down the plant for economic as well as health reasons. We jumped on the opportunity to compare a group of mothers and children born before the closure of the plant to a similar group of children conceived after plant closure. As we had theorized, the second cohort had more favorable birth and cognitive outcomes, significantly lower levels of PAH-DNA adducts in cord blood, higher levels of a protein important in early brain development known as brain-derived neurotrophic factor (BDNF), and longer telomeres (caps on the ends of chromosomes) that are a general marker of health. That was good news, at both molecular and clinical levels.[15]

Attention Problems and ADHD

The combustion-related air pollutants—particulate matter black carbon, and nitrogen dioxide—have been linked to attention problems and ADHD in childhood. Studies in South Korea and Japan reported more attentional problems in children who had higher prenatal exposure to particulate matter. Often the effects of air pollution have varied by sex of the child. The Boston study mentioned above found more attention problems in boys who had experienced higher prenatal exposure to $PM_{2.5}$, but not in girls.[16]

Returning to our NYC cohort, prenatal exposure to PAHs measured by personal air-monitoring or PAH-DNA adducts in cord or maternal blood was associated with symptoms of inattention and ADHD when children were tested at ages 6–9. We found evidence that socioeconomic disadvantage magnifies the harm from pollution: the children with high prenatal PAH exposure (high adducts in cord blood) had more symptoms of ADHD compared to those with low PAH exposure, but the greatest difference was seen among the children whose mothers reported having also experienced the stress of material hardship from pregnancy through their child's early years. Clearly, interventions to protect pregnant women and children must address both of these "toxic" exposures.[17]

Autism

Research during the past 10 years has shone a spotlight on the likely role of air pollution in autism spectrum disorder, a condition defined by social and communication difficulties and repetitive behaviors. Boys are four times more likely to be diagnosed with autism than girls. Studies in the United States and other countries have reported associations between prenatal exposure to $PM_{2.5}$ or traffic-related air pollutants and autistic traits, indicating that these exposures are likely to be causally related to the condition. Postnatal exposure to air pollution has also been implicated as a contributor by research in Israel and Denmark as well as in the United States.[18]

In our NYC cohort, testing of the children at age 11 found a link between prenatal PAH exposure measured by PAH-DNA adducts in mothers' blood at delivery and deficits in social communication (an autistic trait), as well as reductions in children's capacity to regulate their emotions. The effect of PAHs on emotional regulation appeared to be driving the problems in social communication. Here again, a molecular epidemiologic study (using a biomarker to reflect exposure) was able to cast light on a mechanism involved in the harm of PAH exposure to the developing brain.[19]

Mental Health Disorders

Mental health disorders are afflicting a record number of children and adolescents, at a 15% prevalence globally. In the United States, almost 17% of children (1 in 6) are diagnosed with at least one disorder. Researchers have recently begun to focus on the role of air pollution in the mental health of children and adolescents. In a London-based cohort, children who were exposed to higher levels of outdoor air pollution had an increased odds of major depressive disorder at age 18. In the United States, follow-up of a cohort of children in Cincinnati, Ohio through age 12 found that traffic-related air pollutant exposure was associated with self-reported depression and anxiety symptoms in the children. In our NYC study, higher prenatal PAH exposure was also linked to more anxiety and depression symptoms in childhood. Our parallel study in Poland observed that prenatal PAH exposure and maternal psychological distress (a response to stress) interacted to increase the risk of anxiety and depression symptoms in children.[20]

Brain Changes

Extending the groundbreaking early work in opening the black box of brain development, researchers have used magnetic resonance imaging (MRI) to study the impact of prenatal or postnatal exposure to air pollutants on brain development and understand the mechanisms involved. They have shown that exposure to PAHs or fine particulate matter ($PM_{2.5}$) prenatally and in childhood can change the architecture of the brain. In some cases, the observed changes were linked to worse neurodevelopmental outcomes. Our early study in NYC using MRI brain imaging in a sample of 7- to 9-year-old children showed significant correlations between their prenatal PAH exposure and decreased white matter volumes in certain regions of the brain that, in turn, correlated with various cognitive and behavioral problems in the children. (Long thought to be passive, white matter plays an active role in learning and brain functions, providing essential connectivity between different brain regions.) A subsequent report involving a larger cohort in Barcelona, Spain, showed that prenatal exposure to $PM_{2.5}$ during fetal life was linked to a thinner cortex (the area covering much of the brain) in several brain regions when children were imaged at 6–10 years of age; these structural changes partly explained the association they saw between prenatal exposure to fine particles and impaired ability to inhibit inappropriate impulses. Yet another study in Barcelona found that prenatal $PM_{2.5}$ exposure, even at relatively low levels, was associated with a decrease in the volume of a key connective structure in the brains of children who were imaged at 8–12 years: this large nerve tract (the corpus callosum) connects the two brain hemispheres and carries information received in one hemisphere over to the other. As the researchers expected, this specific structural change was linked with a higher hyperactivity score in the same children. Other studies using imaging techniques have found the fingerprints of postnatal exposure to air pollution on the developing brain. These studies have provided critical links in the chain of evidence connecting fossil fuel–generated air pollution to the neurodevelopmental disorders we see more and more frequently in our children and adolescents.[21]

Long-Term Effects of Air Pollution on the Brain

Fortunately, some of these early neurodevelopmental disorders can remit or attenuate as children pass through adolescence and into adulthood, but, like adverse birth outcomes, the effects of toxic exposures on children's cognitive and behavioral functioning often persist into adulthood. At least a third of children with ADHD and half of those with autism continue to have poor outcomes as adults. Another example of such lifelong consequences is the loss of lifetime earnings as a result of IQ reduction. Based on data from our prospective study in NYC showing IQ loss in children who experienced high prenatal exposure to airborne PAHs, we estimated the benefit to NYC children in low-income families if ambient PAH concentrations were reduced by 25%: the estimated gain in lifetime earnings due to IQ increase *for each year's birth cohort* was US $215 million.[22]

As if the effects of exposure on the developing brain were not concerning enough, MRI-based research and epidemiology is now revealing that chronic exposure to airborne pollutants in the early years may "seed" neurodegenerative diseases appearing at older ages. Using molecular biomarkers and MRI of the brain, studies of children living in high air-pollution areas in Mexico City have reported distinct brain changes that are the same as those associated with Alzheimer's disease and mood disorders in adults. Large epidemiological studies have reported greater risk of dementia, Alzheimer's, and Parkinson's disease in older adults due to prior air pollution exposure. Follow-up of almost 1,000 older women in the United States using MRI found evidence that long-term exposure to $PM_{2.5}$ was a risk factor for structural brain changes indicative of increased Alzheimer's disease risk and early memory decline.[23]

Adding up all this evidence, it is clear that air pollution is contributing to adverse neurodevelopmental outcomes, including when exposure occurs during the prenatal period. Air pollution is clearly detrimental to children's cognitive and behavioral development and mental health, often with lifelong consequences for their ability to learn, earn, and contribute to society.

Asthma and Other Respiratory Illness

As with the developing brain, the respiratory tract is highly vulnerable to toxic air pollutants in the first days, months, and years of life. Asthma can appear at any stage throughout life, but it generally develops in childhood. In childhood asthma, the lungs and airways become easily inflamed when exposed to certain triggers, such as air pollutants, pollen, or a respiratory infection. Childhood asthma can cause daily symptoms that interfere with play, sports, school, and sleep. In some children, unmanaged asthma can cause dangerous, sometimes fatal, asthma attacks. Globally the prevalence of childhood asthma is high, about 13%, and the rate has increased significantly in recent decades. As with neurodevelopmental disorders, this increase is only partly explained by improved awareness and better diagnosis. In the United States, asthma rates are markedly higher in communities of color within urban areas. The Harlem Children's Zone project, which operates in a 60-block area in Central Harlem, NYC, screened almost 2,000 children 10–12 years of age and reported that nearly one in three of these children under the age of 13 had asthma or asthma-like symptoms. We have seen a similar rate in our NYC cohort in Northern Manhattan and the South Bronx, where up to 30% of children have had a reported diagnosis of asthma.[24]

Two siblings in the Harlem Children's Zone project, Angela and Willie Vasquez, describe their symptoms differently: "Before it happens I get itchy all over," says Angela, 8 years old. "Then I cough. A lot. And then I can't breathe. I try to breathe in, but I can't. My chest feels real tight, like it was squeezed."

"It happens sometimes after I wake up in the morning," says Willie, Angela's 4-year-old brother. "I'm wheezing, wheezing, wheezing. When I breathe in, it's not enough air." Despite those differences, Angela and Willie agree, emphatically, about the effects of their symptoms. "It hurts," says Angela. "And it's scary."

"It hurts,' says Willie. "And it's really, really, really scary."[25]

The high prevalence of childhood asthma in Harlem and in Northern Manhattan is consistent with reports from other low-income urban communities, where it is common to see a child using an inhaler to deliver medication that relieves lung spasms during an asthma attack.

Particulate matter has long been established as a trigger of severe symptoms and asthma attacks in children who have the disease, causing visits to emergency departments and hospitalization. Nitrogen dioxide and ozone, environmental tobacco smoke, and pollen are known triggers as well. We now know that air pollution is not only able to precipitate asthma attacks in children who have the disease but that it also can act to initiate the disease. Globally, in 2019, an estimated 1.85 million new pediatric asthma cases were attributable to nitrogen dioxide, a transportation-related air pollutant.[26]

Other respiratory outcomes of serious concern associated with air pollution are impaired lung function (very simply, how well the lungs work to move air in and out) and abnormal patterns of lung function growth. Study after study has linked exposure to air pollutants in childhood to both outcomes. For example, a prospective study of almost 2,000 school children in Southern California found that children exposed to higher levels of air pollution, including nitrogen dioxide and $PM_{2.5}$, had significantly lower growth of their lung function at age 18, an age when the lungs are nearly mature and lung function deficits are unlikely to be reversed. On the good news side: when children moved to cleaner areas, the picture improved. In a landmark study, 10-year-old children who relocated to areas with differing levels of particulate matter were tested for lung function 5 years later: those who had moved to areas with lower particulate levels had improved lung function growth rates.[27]

Yet another consequence of air pollution exposure is acute respiratory infection. A study of more than 140,000 children in Utah, mostly under 2 years of age, found that a short-term increase in $PM_{2.5}$ was associated with the development of acute lower respiratory infections in these children. A more recent concern is that, although children have contracted coronavirus at far lower rates than adults and few have died, there is evidence that children who have asthma and live near industrial polluters face higher risk from COVID-19.[28]

As with neurodevelopmental outcomes related to air pollution, these respiratory conditions frequently persist over the lifetime. Data from a study in Melbourne, Australia, showed that half of individuals who had suffered from persistent asthma in childhood and 82% of those who had been classified as having had severe childhood asthma still had asthma symptoms at age 50. In addition, children with severe or persistent asthma are at increased risk of developing permanent airflow obstruction and chronic obstructive pulmonary disease (COPD), a condition that includes chronic bronchitis and emphysema, as adults.[29]

Immune Effects

Air pollution can interfere with the functioning of the immune system, which develops over the entire course of the fetal period and remains immature for many years after birth. What follows can often be the harmful conjunction of damage to key organs and a weakened immune system. Let's take respiratory infections for example: in addition to causing direct harm to the developing lung, exposure to particulate matter, PAHs, diesel particles, and nitrogen dioxide can suppress the ability of the immune system to ward off bacterial and viral infections like respiratory syncytial virus (RSV) and pneumonia. Most of the time RSV will cause a mild, cold-like illness, but it can also cause severe illness such as bronchiolitis (inflammation of the small airways in the lung) and pneumonia (infection of the lungs). In the United States, it is estimated that 58,000 young children are hospitalized due to RSV infection every year. Premature infants are at greatest risk.[30]

We have another vivid example of the role of damage to the immune system in childhood asthma. An elegant study by Stanford researchers has traced the molecular pathway between air pollution, immune system dysregulation, and asthma severity. The study included 181 children with and without asthma living in Fresno and Palo Alto, California. Based on daily air quality data from California Air Resources Board monitoring stations, the researchers calculated each child's annual average exposure to PAHs. The annual average exposure to PAHs was seven times higher for the children in Fresno compared with the children in Palo Alto. Levels of particulate matter were also significantly higher in Fresno. The Fresno group had more severe symptoms of asthma than the children in Palo Alto. Those findings would have been important in themselves, but the researchers also collected blood samples from the children and measured the level of key immune cells—the T-regulatory (T-reg) cells—a subpopulation of white blood cells whose job is to suppress the harmful inflammatory responses that are the hallmark of asthma and other allergic diseases. They found that the children in Fresno had lower overall levels of T-reg function. Delving further into this pathway, the researchers correlated exposure to airborne PAHs with methylation of a specific gene, peculiarly named the Forkhead box transcription factor (Foxp3), that is expressed in T-reg cells. When active, the gene regulates the development and function of these cells, thus turning down the inflammatory response. The researchers showed that PAHs caused DNA methylation in Foxp3, effectively switching it off, disabling the gene's function, and increasing the severity of asthma symptoms. This is one of few studies that has linked increased ambient air pollution exposure to dysregulation of a specific immune cell, then to inactivation of a specific gene involved in that cell's function, and, finally, to the severity of asthma symptoms in the children—in the words of Kari Nadeau, the lead author, from "A to Z."[31]

In addition to the many health outcomes described above, recent research has associated sleep disorders with exposure to particulate matter, both when children are exposed prenatally and when exposure occurs in childhood. This research adds yet another concern about the impacts of air pollution on children's health.[32]

HEALTH IMPACTS OF CLIMATE CHANGE

Climate change is a threat multiplier for infants and children. It has already taken a significant toll on children, and the toll is increasing every year. The health effects are generally considered under the headings *direct* and *indirect*, reflecting how closely they are linked to climate change. For example, increased temperature due to climate change directly affects children's health through severe heat events and indirectly through major floods, intense storms, and extreme forest fires—all of which are occurring more frequently due to climate change. Other indirect effects include malnutrition and undernutrition due to food insecurity prompted by drought, the spread of infectious-disease vectors, respiratory illness due to increased air pollution and aeroallergens, and mental health disorders from displacement and social and political instability. However, the labeling doesn't much matter to children who are especially vulnerable to both the indirect and direct consequences of climate change.[33]

There are no estimates of the totality of harm to children from climate change. The estimates that exist are partial, usually focused on one or a few health damages (such as deaths, diarrhea, malaria, malnutrition). But the true health burden is far greater. The latest report is that, right now, almost every child on earth is exposed to at least one climate and environmental hazard, shock, or stress, and 1 billion children live in countries that are at an extremely high risk from the impacts of climate change.[34]

Today, as before, children bearing the greatest burden of climate-sensitive diseases are those living in low- and middle-income countries with the least capacity to adapt to risks—regions that have contributed the least in terms of global emissions of greenhouse gases. However, although children in these countries are most at risk, the impacts of climate change are increasingly being seen in the United States and Europe, especially among populations of low socioeconomic status and communities of color.

Just as there are no comprehensive assessments of the health burden falling on children, there are no assessments of the full economic toll of child ill health and developmental impairment on families and society due to climate change. However, we can gauge the magnitude of the economic cost from the estimate that, in the United States alone, the health costs of fossil fuel–related air pollution and climate change far exceed $800 billion a year.[35]

Weather Disasters: Floods and Severe Storms

Climate change has intensified major floods and hurricanes that have caused drowning, physical injury, and traumatic stress in children.

An estimated 600,000 people died in the 1990s as a result of weather-related "natural" disasters, of whom many, if not most, were children. A later estimate held that up to 175 million children were likely to be affected every year by "natural" disasters attributed to climate change. Although no individual severe storm or single major flood can be attributed directly to climate change, there is overwhelming agreement among scientists that the continuing patterns of global warming are contributing to the intensity and frequency of such events. Since the 1990s, climate-induced severe weather disasters have risen by an average of 35% globally. Whether called "hurricanes" in the Atlantic Ocean, "typhoons" in the western Pacific Ocean, or "cyclones" in the Indian Ocean, tropical cyclones have become more intense. The marked regional and global increase in the proportion of the strongest hurricanes (category 4 and 5 storms) has been directly attributed to human-caused global heating of the climate.[36]

We think first of the danger from high winds, but the greatest damage to life and property is from secondary connected events such as storm surges, flooding, landslides, and tornadoes. The high winds from tropical storms push massive surges onto land, compounding the effect of sea level rise on coastal and inland flooding. In fact, of all the natural disasters, floods are the most common, accounting for about 40%. In addition to coastal flooding, climate change has increased the risk of flash floods in rivers through extreme precipitation events, colloquially termed "rain bombs." By 2015, more than half a billion children were identified as living in areas with extremely high levels of flood occurrence.[37]

Drowning accounts for 75% of deaths in flood disasters, most of which are in children. Risks are greatest in low- and middle-income countries where people live in flood-prone areas and where systems to warn, evacuate, and protect communities are nonexistent or only just developing. Take the massive flooding across Southeast Asia in 2011: an estimated 9.6 million people were affected; more than 200 children died in Cambodia, Vietnam, and Thailand; and thousands of schools were damaged. In 2013, largely as a result of the massive storm surge from Typhoon Haiyan, known in the Philippines as Super Typhoon Yolanda, more than 6,000 people died in that country alone and nearly 6 million children were affected. Then, in 2017, massive flooding in South Asia placed almost 16 million children and their families in urgent need of life-saving support.[38]

The impacts of climate-related natural disasters are felt by children worldwide. In 2005, Hurricane Katrina hit New Orleans, Louisiana, causing over 1,800 deaths, forcing 1 million people from their homes, and leaving more than 370,000 children without schools. In 2012, Hurricane Sandy affected people in eight countries, with particularly severe damage in New Jersey and New York, where the schools of more than 20,000 students were closed and more than 600,000 housing units were destroyed. In 2017, Hurricane Harvey caused the deaths of 107 people in Texas, including a number of children, and affected up to 3 million children and their families. Most recently, the remnants of Hurricane Ida caused widespread flooding from Louisiana to New England and dozens of deaths from drowning, including a 2-year-old toddler.[39]

In addition to physical harm, children are highly vulnerable to trauma and stress from the experience of extreme weather events, increasing the likelihood of mental health problems. Approximately 100 million youths globally are exposed to disasters each year: as many as 72% of them show posttraumatic stress symptoms when surveyed during the first 3 months after a disaster. Posttraumatic stress symptoms in the young are associated with poorer mental and physical health, worse academic performance, and later employment problems. Children who were directly affected by Hurricane Katrina were later found to have difficulties concentrating in school, more behavioral problems, greater anxiety, and other persistent mental health issues. More than 60% of 200 New Orleans schoolchildren who were screened 15 months after the hurricane had high rates of depression, anxiety, and posttraumatic stress disorder (PTSD). Another school-based study indicated that 41% of children directly exposed to the hurricane were in need of mental health services nearly 2 years later.[40]

One such example of the pervasive effects on children is the story of then 5-year-old Sheniya Green and her 3-year-old sister. A published account of this child's ordeal describes how, for 8 hours during that night in August 2005, she and six of her relatives clung to the roof of her grandfather's house in New Orleans' Lower Ninth Ward while the flood waters surged around them. During the night, Sheniya fell into the flood waters but was rescued. Sheniya's grandmother and 3-year-old sister died. Sheniya's family relocated to Houston, Texas, trying to rebuild their lives while mourning their losses. A year later Sheniya was still expecting her little sister to come home and experiencing psychological trauma from the loss.[41]

It now appears that we have to add severe winter storms to our list of climate change impacts. Although global climate change is clearly causing warmer winters and record hot temperatures, counterintuitively, it may also be partly responsible for extreme winter storms in the mid-latitudes of Europe and the United States. As mentioned in chapter one, the rapid heating of the Arctic (at a rate more than twice the global average) has caused the jet stream to shift, allowing frigid air normally concentrated around the north pole in the polar vortex to reach farther south. In 2021, this phenomenon contributed to huge snowstorms in Europe as well as in parts of the United States normally accustomed to milder winters. In February of 2021, a record-setting winter storm brought ice, snow, and freezing temperatures across central and southern parts of the United States. Texas was most affected, as extreme weather collided with a lack of climate change resilience: at least 57 people died, including several children who died from hypothermia. Hundreds of thousands were without electricity, millions without clean water; there were food shortages and schools were closed.[42]

Heat

We are all keenly aware that, over the past several decades, there have been more and more heatwaves (2 or more days of unusually hot weather), breaking temperature records. As we saw in the previous chapter, pregnant women, developing fetuses, and young children are especially vulnerable to heat. Because of climate change, heatwaves have become both hotter and longer across the United States, Europe, and Asia. Globally, 2020 was the hottest year on record: temperatures topped 120°F around the world—in India, Pakistan, and the United States. According to a recent analysis, without rapid global action to cut global emissions, children born in 2020 will experience two to seven times more extreme heatwaves on average in their lifetimes than people born in 1960.[43]

Higher temperatures are a risk factor for adverse birth outcomes through dehydration or by impairing fetal growth. We saw earlier that large-scale studies in the United States and China have found that mother's exposure to heat is associated with both preterm birth and low birth weight and that heatwaves and $PM_{2.5}$ may interact synergistically to increase the risk of preterm birth.[44]

Further direct effects of heatwaves are infant deaths and hyperthermia (elevated body temperature), heat stress, kidney disease, and impacts on mental health of children. Infant deaths have been increased during heat waves in Europe, China, Australia, and many other countries. As the planet has become warmer, the incidence of heat-related illness in children is increasing. Almost half of heat-related illness are in infants and children. During a deadly heat wave in California in 2006, children ages 0–4 years and the elderly were at greatest risk of heat-related illness. Heat-related deaths have occurred in infants and young children left unattended in vehicles or excessively warm environments and in exercising adolescents. Heat illness is among the leading causes of death in young athletes, and the rate has been increasing. Between 2016 and 2018, nearly 12% of US emergency room visits among children 18 and younger were due to heat.[45]

Subtler, less recognized effects of extreme heat during fetal development, infancy, or childhood are reduced cognition and lower lifetime earnings. Each additional day a fetus or infant experiences 90-plus-degree (32°C) temperatures, he or she makes an estimated $30 less a year on average, or $430 over the course of the lifetime. Unless countries take strong action on climate change, by the end of this century there will be an estimated 43 such days a year, boosting the average earning loss to $1,290 per individual. On a population basis that is huge. Possible explanations include cognitive impairment due to preterm birth or lower birth weight, poor academic performance during high heat days, and heat-related illness that causes children to miss school.[46]

Warmer temperatures act in yet another way to affect children's health: they increase exposures to toxic air pollutants, chemicals, and infectious agents. Higher temperatures in areas with low rainfall cause volatilization of toxic chemicals and pesticides to which children's developing nervous systems are particularly vulnerable. Higher temperatures also cause migration of some pests and insects to more northern, cooler regions, resulting in more infectious disease and greater use of neurotoxic pesticides—another vicious circle.[47]

Wildfire and Smoke

More extreme forest and wildland fires are another effect of climate change directly affecting children's health. In the Western United States, Indonesia, Australia, the Amazon, Mediterranean Europe, and the Arctic Circle, higher temperatures have increased the frequency of extreme forest fires or "megafires." Warming produces yet more warming since the fires destroy a natural buffer for climate change at the very same time as they emit billions of tons of CO_2, methane, and other greenhouse gases every year. Globally, fires may contribute as much as 20% of total carbon emissions. (These emissions are nonetheless dwarfed by those from the burning of coal, oil, and gas.) Forest fires also inject vast quantities of particulate matter, black carbon, hydrocarbons, and volatile organic chemicals into the atmosphere in the form of aerosols (fine particles and droplets) and gases. Ozone is formed from several of these precursors and more rapidly at higher temperatures. Because of their small size, these pollutants can be transported thousands of miles from their origin and affect health in large swaths of the population. For example, wildfires in Canada in 2017 resulted in extreme levels of aerosols over Europe, higher even than those measured after the cataclysmic eruption in 1991 of the Mt. Pinatubo volcano in the Philippines.[48]

Over the past decade, the United States has seen an average of 70,000 wildland fires every year, burning an annual average of 7 million acres. The Southeast, Pacific Northwest, and California have seen dramatic increases in forest fires in recent years. The contribution of wildfire smoke to $PM_{2.5}$ concentrations in the United States has more than doubled since the mid-2000s, in recent years accounting for an estimated 25% of $PM_{2.5}$ across the United States and up to half in some Western regions. There were an estimated 7.4 million children in the United States affected by lung-damaging wildfire smoke every year between 2008 and 2012. Not only do children breathe more air relative to their body weight than adults, but they have less nasal deposition of particles, so a higher proportion of particles winds up in the lower lung. Compared to some other sources of air pollution, smoke from forest fires contains higher levels of PAHs, volatile organic chemicals, and other ozone precursors, which means more potential to create harmful oxidative stress and inflammation in developing brains and lungs.[49]

Wildfire smoke affects the young in many ways. Exposure to smoke in pregnancy has been linked to decreased birth weight and preterm birth. Respiratory effects head the list of pediatric illness due to exposure to forest fire smoke. In the United States, the estimated number of emergency room visits in children with asthma due specifically to ozone generated by wildfire emissions exceeds 2,000 visits every year. Other respiratory illnesses, including wheezing, pneumonia, and bronchitis, are also increased during wildfires. Exposure of young lungs to intense episodes of wildfire pollution may trigger the development of asthma, as we saw with fine particulate matter from other sources. It is not just children living close to the fires who suffer; millions of children living far away are exposed to wildfire smoke that has traveled hundreds of miles on prevailing winds. Although diluted by travel, the pollution can worsen existing respiratory problems like asthma.[50]

A new concern has recently emerged about the health effects of forest fires: in the summer of 2020, as the Western United States faced an already active wildfire season, public health officials warned that smoke exposure was likely to worsen the coronavirus pandemic by increasing susceptibility to the virus. Children make up a small percent of Coronavirus cases (9% in the US) and the disease is generally milder than in adults. However, between March 1, 2020 and mid-August 2020, 576 US children required hospitalization for COVID-19. Black and Latino children were disproportionately affected because of underlying health conditions and co-exposure to air pollution.[51]

As with hurricanes, wildfire disasters cause serious mental health effects in children. Eighteen months after a large wildfire caused evacuation of a small Canadian city, middle and high school students showed elevated rates of depressive symptoms compared to youth in a control city. Wildfire smoke and the trauma of evacuation likely combined to affect the children's mental health. Similarly, studies of youth affected by wildfires in California and Australia reported high levels of stress and symptoms of PTSD after the fires.[52]

Drought and Malnutrition

Droughts have increased worldwide in frequency, duration, and intensity at an unprecedented rate due mostly to climate change. The result has been widespread crop failure, livestock deaths, and massive food insecurity, disproportionately affecting children in poor families. More than 11 million people have died and more than 2 billion have been affected worldwide by droughts since the turn of the century—more than any other physical hazard. The regions affected include Central and South America, Australia, Sub-Saharan Africa, Afghanistan, other parts of Asia, and the Southeastern United States. In 2015, almost 160 million children lived in areas of high or extremely high drought severity. Globally, the number of people exposed to droughts could increase by 9–17% in 2030 and by as much as 50–90% in 2080.[53]

With such massive food insecurity has come a sharp increase in malnutrition in children. In 2018, Guatemala recorded more than 15,000 cases of acute malnutrition in children under 5, as drought linked to climate change reduced food harvests. (Figure 3.1 shows 1-year-old Aníbal who is recovering from malnutrition in Guatemala.) We saw in the previous chapter that the greater nutritional needs of infants and children place them at greatest risk from famine. Malnutrition during the first 1,000 days of life causes stunting of the brain and body, with associated behavioral and cognitive problems. The result is that children are less able to learn and will have lower lifetime earnings. In 2020, globally, 149 million children under the age of 5 years (22% of children in that age group) were stunted, another 14 million suffered from acute malnutrition. Malnutrition increases children's susceptibility to infection, and infection contributes to malnutrition, setting up yet another vicious cycle (Figure 3.1).[54]

We tend to think of drought as a problem in developing countries, but in the fall of 2020, more than 45% of the continental United States was in drought; in the spring of 2021, the same conditions prevailed. Though only about 1% of children in the United States suffer from chronic malnutrition, many children are at risk of undernutrition. In a recent year, 13 million US children lived in food-insecure homes and were vulnerable to undernutrition. That number increased to almost 17 million US children during the COVID-19 pandemic as parents lost jobs.[55]

Although drought is the most prominent and recognizable cause of malnutrition in children, other impacts of climate change—higher temperatures, water scarcity, floods, and changing patterns of agricultural pests and diseases—have also increased food insecurity around the world.

Figure 3.1 In Guatemala, 1-year-old Aníbal is recovering from malnutrition.
Credit: Coopi Internazionale, 2019.

Compounding the problem, greater CO_2 concentrations in the atmosphere are causing the decline of nutritional quality in food crops, reducing protein content and essential minerals like zinc.[56]

Infectious Disease

Climate change is increasing the risk of infectious disease in children in many regions. Malaria, Dengue fever, Zika, and tick-borne diseases such as Lyme disease take advantage of immature immune systems to disproportionately affect children. Climate change has had the effect of changing the duration of the transmission season and geographic spread of the species that carry these infectious agents. The geographic range of *Anopheles* mosquitoes that carry malaria (caused by a parasite) has changed in response to climate change. As a result, the incidence of malaria in some regions, such as highland areas of Ethiopia and Colombia, has increased due to warmer temperatures, with ominous implications for the malaria burden in the densely populated highlands of Africa and South America. Dengue is the most important mosquito-borne viral disease globally, impacting half the world population—in this case, it is the *Aedes* mosquito that is the carrier. The increase of temperature has been linked to an expansion of dengue transmission at higher altitudes, resulting in a rise in dengue incidence in the mountainous country of Nepal. Although dengue is most common in tropical regions, with a warmer climate, mosquito vectors are now present for a greater part of the year in North America, so the incidence of dengue is expected to increase there as well. In some areas, such as in Sub-Saharan Africa, however, where climate change is expected to decrease rainfall, the number of mosquitoes may decrease, and so malaria transmission rates would also fall. The *Aedes* mosquito is also the primary carrier for the Zika virus. From 2007 to 2016, the virus spread eastward, across the Pacific Ocean to the Americas, leading to the 2015–2016 Zika virus epidemic. In a first for the United States, locally transmitted cases occurred in Florida and Texas. The pattern of Lyme disease has also changed. Traditionally more prevalent in the Northeastern United States, Lyme disease has now been found in all 50 states and the District of Columbia. Climate change is not the only factor in these dramatic shifts in the pattern of infectious agents (changes in land use, population density, and human behavior also contribute), but it plays an important role.[57]

The toll on children's health from these infectious agents is huge and growing. Malaria causes substantial illness and death in children in many of the most resource-limited areas of the world. Children under 5 years of age are one of most vulnerable groups. In 2018, over 270,000 children died of malaria before their fifth birthdays. The vast majority of dengue cases are children of less than 15 years of age. The Zika virus uniquely affects the developing fetus. Transmitted from a pregnant woman to her baby, the virus evades fetal innate immune responses to move into the fetal brain, causing microcephaly (shorter than normal head), severe brain malformations, and other birth defects.[58]

Of the estimated 476,000 people diagnosed with Lyme disease in the United States every year, the highest reported incidence occurs in children ages 5–14. If Lyme disease is caught and treated early, most children will make a full recovery. However, some children with Lyme disease go on to experience what is called a *post-infectious syndrome*, with symptoms such as fatigue, joint aches and pains, headaches, difficulty sleeping, and problems concentrating.[59]

Children are also more susceptible than adults to cholera and other infectious diarrheal diseases caused by crop and water contamination from storms and floods. Repeated diarrheal infections in childhood can affect children's education and cognitive development, negatively impacting a family's economic situation, which leads in turn to poor health. The impacts are not only felt in developing countries. Salmonella, a food-borne infectious disease, has become more prevalent across much of continental Europe as a result of higher temperatures.

Allergy and Asthma

As a result of climate change, we are seeing—and many of us are experiencing—more allergy and respiratory illness in children from increased pollen production by plants at higher temperatures. The fertilizing effects of CO_2 and higher temperatures have both contributed to increased growth, reproduction, and production of pollens from certain trees, grasses, and weeds. The increase in pollen levels across North America since 1990 has been dramatic. This trend is strongly coupled to observed warming and parallels the increase in pollen sensitization in children. It also aligns with the rising trend in the numbers of adolescents and adults with allergic asthma. In addition to triggering symptoms of asthma, hay fever, and eczema in children with allergies, high-pollen periods are associated with greater susceptibility to respiratory infections by intensifying lung inflammation and weakening immune responses.

Interactions occur here as well. Air pollutants like ozone, particulate matter, and sulfur dioxide can collaborate with pollen in several ways to increase risk: their inflammatory effects make it easier for pollen allergens to penetrate into the airways; the air pollutants promote the release from pollen grains of the antigens that lead to allergic responses; and the pollutants are able to absorb pollen grains and thus prolong their retention in the body.

Mold is another trigger of asthma attacks, in this case whether the child is allergic or not. Excess moisture from heavy rains and flooding due to tropical cyclones, hurricanes, and thunderstorms can lead to proliferation of mold in homes. In the aftermath of Hurricane Katrina, of 112 households in the New Orleans area assessed by the Center for Disease Control, almost half the homes had "visible mold growth" and 17% had "heavy mold coverage."[60]

Forced Migration and Displacement

Surprising some of his conservative followers, Pope Frances wrote in 2016 that climate change is contributing to the "heart-rending" refugee crisis. Worldwide, more than half of internationally displaced people have been children. Here, like the Pope, I use the term "refugee" broadly to encompass people who have fled persecution and armed conflicts (legally recognized "refugees"), people who have applied for refugee status and have legal protection ("asylum seekers"), and migrants who were forced by unbearable conditions such as drought and natural disasters to leave their homes ("climate migrants" or "climate refugees" who have no legal status). As severe climate change displaces more people, I anticipate that the international community will be forced to legally expand the term "refugees" to include "climate refugees."

The effects of climate change have become even more important "push factors" in migration than economic inequality or conflict. In fact, up to three times as many people are being internally displaced every year due to natural disasters as from armed conflicts or other forms of violence, and much of what is now international migration started out as weather-related internal displacement. Climate-related events have already contributed to over 50 million children being forced from their homes. Globally, each year more than 20 million people are forcibly displaced by weather-related disasters such as floods, storms, wildfires, and extreme temperature. These climate refugees now account for more than a half of all migrants. Forecasts vary from 25 million to 1 billion climate migrants by 2050, moving either within their countries or across borders. Even the modest estimate of 200 million is staggering. Imagine what that influx would do to political and social stability in Western countries where immigration has deepened political polarization: it is cited as one of the biggest factors in the UK's referendum on membership in the European Union (Brexit) and has played a major role in recent elections in Germany and France.[61]

Think, too, what this would mean to children, who will undoubtedly continue to represent half of climate refugees. Children are most at risk from displacement physically and emotionally: they are prone to physical injury, chronic stress, and psychological trauma as a result of being forced—with their families and sometimes alone—to flee their homes. The mental health consequences include anxiety, depression, and posttraumatic stress. In developing countries, when families are forced from their homes, children often become separated from their parents, increasing their susceptibility to violence, exploitation, and abuse. Although the burden is especially great in low-income, developing countries, once again, the United States and other rich countries are not spared. More than 1 million displacements have occurred on average in recent years in the United States as a result of disasters. That figure is likely to grow as climate change increases the severity of wildfires, hurricanes, and coastal flooding. Heat and extreme weather fueled by rising carbon emissions may force 1 in 12 Americans in the Southern half of the United States to migrate north over the next 45 years. This would mean a displacement of more than 20 million people, including millions of children.[62]

In another of the many vicious cycles we encounter with climate change, migration can be triggered by conflicts, but migration due to the scarcity of food or extreme weather events can itself trigger conflicts. These climate-related environmental impacts interact with social, economic, and political factors to increase the likelihood of wars and armed conflicts in vulnerable countries. The war in Syria and conflict in the Sahel region of Africa are examples of disasters attributable in part to climate change. These have displaced millions of people, as many as half of them children. In Syria changing weather patterns caused a drought from 2007 to 2010, leading to massive food insecurity that helped fuel the Syrian war, now in its seventh year. The war has resulted in more than 470,000 deaths, over 13 million people in need of humanitarian assistance, almost 7 million people internally displaced, and nearly 5 million people forced to live in camps in Turkey, Jordan, and Lebanon. (Figure 3.2 shows a Syrian child with his mother, living in a refugee camp in Turkey.) More than 3 million Syrian refugees now live in Turkey—the largest refugee population in the world—most in host communities that are themselves poverty-stricken. Almost half of these Syrian refugees are children, and hundreds of thousands of them are out of school. While this devastating war was not solely a result of climate change, drought was a key factor.[63]

Figure 3.2 Syrian refugee children in an informal tent settlement in Lebanon. They are burning plastic and other trash to stay warm.
Credit: Jon Warren, 2016 World Vision.

Disasters and displacement have broken apart families and smashed so-cial norms and supports, propelling an increase in violence against girls and women, rape, and sexual trafficking of children. Women and girls are more likely to be targeted by violence than men, especially those of lower social class and racial/ethnic minorities. In the United States, violence against women soared after Hurricane Katrina among women who were displaced. Around the world, armed conflict, displacement, natural disasters—all of which are linked to climate change—increase the vulnerabilities and desperation that enable trafficking to flourish, with migrant women and children being targeted.

Effects on Mental Health

Among the most striking and worrisome consequences of extreme weather disasters, fires, drought with resultant malnutrition, and displacement are the mental health effects of these childhood experiences. Much is known about the effects of adverse childhood experiences such as abuse, neglect, and household disfunction on brain structure and the develop-ment of key areas of the brain. We may now add to the list climate-related disasters, food scarcity, and displacement. Common forms of mental ill-ness in children who have suffered various adverse childhood experiences include anxiety, depression, aggressive behavior, PTSD, and substance use problems. Adverse experiences not only raise the risk for mental disorders in childhood, but also confer a lasting vulnerability to anxiety, depression, and mood disorders in adulthood. Approximately one-third of all mental disorders worldwide have been attributed to exposure to adverse experiences in childhood.[64]

Even in the absence of direct experience of impacts of climate change, stress due to the awareness of climate change and its impacts is having a significant effect of the mental health of children everywhere in the world. As children reach the fragile state of adolescence and learn, either first-hand or through reports, of the climate threats to their very existence, this knowledge fosters a kind of PTSD. There is now a term, "climate change anxiety" for the condition that many young people are experiencing. Here are some words of young climate activists: "It's really hard to grow up on a planet full of ifs," said Jamie Margolin, the 17-year-old co-founder of Zero Hour and climate activist from Seattle. "There's always been a sense that everything beautiful in this world is temporary for my generation."

"Coming to terms with the fact that we have lived on a dying planet is terrifying," despaired 18-year-old activist Kaylah Brathwaite, a Charlotte, North Carolina, college student. "I'm just going to have to be scared for the rest of my life."

Max Prestigiacomo of Madison, Wisconsin and the Youth Climate Action Team, said: "A lot of us say we can't think more than 16 months ahead because we don't know the environment we will have to grow up with."

"Young people from different parts of the world are living in constant fear and climate anxiety, fearing their future," mourned Komal Kumar, a youth climate activist from Fiji.[65]

Recent data testify to the serious mental health burden being inflicted on young people. In a survey across 10 countries, nearly 60% of young people said they felt very worried or extremely worried about climate change; more than 45% said their feelings about climate change negatively affected their daily lives. Two-thirds reported feeling sad, afraid, and anxious. One 16-year-old said: "It's different for young people—for us, the destruction of the planet is personal."[66]

Cumulative Impacts of Air Pollution and Climate Change

There is a dangerous interaction between toxic air emissions from the burning of fossil fuels and climate change, one leading to greater exposure to these hazards. Higher temperatures due to climate change accelerate the formation of ground-level ozone from its precursors (volatile organic chemicals, carbon monoxide, and nitrogen dioxide). As a result, ozone levels are sharply elevated during heatwaves, as during the summer of 2003, when Europe experienced record high temperatures that resulted in at least 30,000 deaths. Climate change has increased ozone levels over large areas in the United States and Europe, especially in the summer. Worldwide, ozone is responsible for several hundreds of thousands of premature deaths and tens of millions of asthma-related emergency room visits annually, many in children. On high ozone days, there are more school absences, more emergency department visits, and more hospital admissions for childhood asthma. In NYC, every year, 1,800 emergency department visits and more than 400 hospital admissions for asthma in children under 18 are attributable to ozone exposure. Ozone is not only a short-term trigger of respiratory illness, but long-term exposure to ozone also is linked to decreased lung function and abnormal lung development in children. [67]

Another result of the interaction between pollutant emissions and climate change is the increase in concentrations of $PM_{2.5}$ in some areas because of changes in temperature, precipitation frequency, and air stagnation due to climate change. In addition, forest fires made more frequent and severe by climate change release large amounts of particulate matter, PAHs, and black carbon, adding to and potentially interacting with the ambient load from fossil fuel burning.

Exposure to both climate hazards and unsafe air quality is common. An estimated 850 million children—1 in 3—live in areas where at least four climate and environmental shocks overlap. Additive and greater than additive (synergistic) effects are possible. The exposures may be co-occurring and chronic, like air pollution, toxic chemicals, and stress; or they may be intermittent and sequential, as with heatwaves or disasters, each acting as a "hit" that heightens the damage from its successor. Interactions may occur when pregnant women or children have combined exposure to air pollution, extreme temperatures, food insecurity, and/or a severe storm. Combined exposure of infants and young children to neurotoxic air pollutants and malnutrition due to drought can lead to synergistic effects on cognition. The defects in immune function due to air pollution exposure make children highly vulnerable to infectious agents whose reach has spread due to climate change. There is sufficient evidence of a synergistic interaction between heat and air pollution on cardiovascular and respiratory morbidity and mortality in adults and children. More studies are needed with respect to child health outcomes. However, as mentioned earlier, researchers have observed a synergistic effect of prenatal exposure to heat and air pollution on risk of preterm birth.[68]

Of particular concern are the cumulative impacts of air pollution and climate change on the mental health of children. As we have seen, the effects of air pollution and climate change extend beyond the physical and immediate developmental impacts to encompass mental health disorders which are afflicting a growing number of children and adolescents. Researchers are now assessing their cumulative impacts to fully grasp both the mental health costs and their flip side: the benefits of transitioning away from fossil fuel. This is a relatively new area of research, but we already have several examples. Starting before birth, exposure to air pollution and higher temperatures increase the risk for preterm birth; preterm babies are then at increased risk for emotional/behavioral problems, depression, and anxiety. Stress and air pollution exposure experienced by the child can interact to further raise the risk of mental health problems in childhood and adolescence. Once again, "de-siloing" the twin threats from fossil fuel leads us to think about holistic interventions—in this case with respect to mental health disorders in the young.[69]

PETROCHEMICALS AND PLASTICS

Awareness of the harm of air pollutants and the effects of climate change has been building for a while, but there is another threat to children's health, also derived from fossil fuels, which deserves more notice. It is the staggering number and quantity of toxic chemicals derived from primary components of crude oil and gas, called *petrochemicals*. Most industrial chemicals, plastics, synthetic fibers, and resins that are used to make thousands of everyday products are produced from petrochemicals. The products include plastic goods, pesticides, fertilizers, clothing, children's toys, electronics, household products from carpeting to shampoo, digital devices, drugs and medical equipment, detergents, and tires, to name a few.

Toxic chemicals in these products have widely contaminated the air, drinking water, food supply, the home environment, and the bodies of almost every human being on the planet. Biomonitoring studies have detected many dozens of synthetic toxic pollutants in the blood and urine of pregnant women and children and in umbilical cord blood. For all the same reasons that the developing fetus and child are more vulnerable to air pollution and climate change, they are disproportionately affected by toxic chemicals. The 2017 Report of the Lancet Commission on which I served, described the "global pandemic" of neurodevelopmental toxicity attributable to synthetic chemicals. It noted that fewer than half of the 5,000 chemicals that are produced in greatest volume, and to which there is nearly universal exposure, have undergone any testing at all, much less for possible effects on the developing brain. We do, however, know a fair amount about some of the neurotoxic, petrochemical-based chemicals like phthalates, the organophosphate pesticide chlorpyrifos, polybrominated flame retardants (PBDEs), and bisphenol A (BPA). These chemicals are *endocrine disruptors*—able to mimic, block, or alter the actions of normal hormones—and can act at low levels of exposure to affect the developing brain and reproductive system.[70]

Phthalates are a group of chemicals that have been used to make plastics more flexible in hundreds of products from packaging materials and flooring to the coatings of medicines, cosmetics, perfumes, teethers, and soaps. Prenatal exposure to phthalates has been associated with ADHD, other behavioral problems, adverse cognitive development including lower IQ, and impaired social communication, one of the signal traits of autism. In our NYC cohort study begun in 1988, 100% of the mothers were found to have multiple phthalate metabolites in their urine during pregnancy. Follow-up of the cohort into childhood showed that higher prenatal exposure measured by this biomarker was associated with reduced IQ at age 7. Phthalates are also toxic to the developing reproductive system: exposure of pregnant mothers to phthalates has been linked to smaller penis size and incomplete descent of testicles, as well as other feminization-type changes in the genital tract of boys.[71]

Chlorpyrifos and other organophosphate pesticides are found in food as a result of their use on agricultural crops to control insects and other pests that may affect crop growth. These include foods that are frequently consumed by children. Prior to its phase-out in 2001, chlorpyrifos was widely used to control household pests. In our NYC study begun in 1988, chlorpyrifos was detected in about 70% of newborn blood samples. Prenatal exposure to the chemical, indicated by the level in newborn cord blood, was linked to symptoms of ADHD, a sharp reduction in children's working memory, and lower full-scale IQ. Distinct changes in the structure of the children's brains measured using MRI and Parkinson's disease-like symptoms in the children were also seen in the more highly exposed children.[72]

PBDE flame retardants have been used in many applications including children's products, plastics, furniture, upholstery, electrical equipment, and electronic devices. Prior to the phase-out of major PBDEs in 2013, these chemicals could be found in the blood of virtually every American. Following the phase-out, levels in the population came down, but exposure continues. PBDEs are sometimes called "forever chemicals" because they resist degradation and bioaccumulate in the fat tissue of living organisms. Our Center's second prospective study of Asian, White, and Black NYC mothers and newborns, launched in 2001, found that children with higher prenatal exposure to PBDEs, evidenced by concentrations of the chemicals in their cord blood, scored lower on tests of mental and physical development between the ages of 1 and 6 years, including verbal and full IQ scores.[73]

BPA is an industrial petrochemical produced in the millions of tons annually in the United States (6 billion pounds globally) to make epoxy resins that are used in the lining of food cans and hard plastic items such as baby bottles, reusable water bottles, food containers, pitchers, tableware, and storage containers. BPA is implicated in neurodevelopmental and reproductive abnormalities. In our Center's main cohort study of Black and Latina women and children in NYC, higher levels of BPA in mothers' urine collected during pregnancy were associated with more anxiety, and depression in the children, mainly in boys.[74]

Plastic pollution is a veritable plague on the planet. As the clean energy transition threatens the bottom line of the fossil fuel industry, these companies are turning to petrochemicals as a market for oil and gas. And the market is booming. The trends in production of petrochemicals and the demand for the products made from them are sharply up, with the demand for plastic nearly doubling since 2000. Huge new petrochemical plants are being built in the United States, Europe, the Middle East, Asia, and Latin America. In the United States, producers are drastically expanding their footprint in the Gulf Coast of Texas and Louisiana, including in a stretch along the lower Mississippi River known as "Cancer Alley", a region already hard hit by pollution and climate change. The companies are also creating a new plastics corridor in Ohio, Pennsylvania, and West Virginia, anchored by a behemoth of a petrochemical plant being constructed outside Pittsburgh, Pennsylvania. That plant alone would produce almost 2 million tons of plastic and release hundreds of tons of toxic compounds into the air every year, not to mention an estimated 2 million tons of CO_2 annually. If the build-out goes forward as planned, it will lock in more climate pollution, plastic waste, and toxic chemicals that contaminate these communities. Since mass production of plastic began 60 years ago, it has created more than 8 billion metric tons of plastic, of which 91% was never recycled. Much of it winds up in the ocean and on beaches; Figure 3.3 shows plastic waste jamming a beach in Manila, Philippines. Without significant action, there may be more plastic than fish in the ocean, by weight, by 2050.The degradation products of plastics (nano-and micro-plastics) have

Figure 3.3 Waste on a beach in Manila, Philippines.
Credit: Photo by Jes Aznar/Getty Images.

been found in placenta and biological samples from newborns and infants, but their potential health effects are largely unexplored.[75]

It doesn't have to be like this. An alternative path for the chemical sector has been proposed that will be described in Chapter 7, "Solutions Now."

Impacts Throughout the Life Course and Beyond

Prenatal or childhood exposure to toxic air pollutants, the impacts of climate change, and toxic chemicals can impair children's health for the rest of their lives. They may do this by launching a trajectory of ill health and impairment in which the initial adverse effect persists as such or by "seeding" latent disease that only becomes evident in later life. Earlier, we saw examples of lower birth weight, preterm birth, cognitive and behavioral problems, respiratory illness, stunting of children's bodies and brains, and mental health disorders that can have life-long consequences, adversely affecting professional and personal life and increasing healthcare costs for individuals and families. Loss of IQ is linked to a reduction in lifetime earnings, in turn affecting the socioeconomic status of the next generation and perpetuating poverty. The psychological and emotional impacts of climate change on children not only encompass the acute, traumatic effects of extreme weather events, forest fires, and forced migration; they also include long-term mental health problems resulting from direct experience to or anxiety about future risks. As if this all wasn't concerning enough, there is growing evidence of possible transgenerational impacts of early-life exposures to air pollutants, toxic chemicals, nutritional deprivation, and stress, possibly via the transmission of epigenetic changes.[76]

Adding all this up, in order to fully account for the harm of fossil fuel, we must take a long view of the lives of our children and theirs.

CONCLUSION

Though daunting, it is past time for us to consider these harms to children in a holistic fashion; it is only by doing so that we can take the proper actions in our communities and advocate with the intensity needed to make change. The vast body of research reviewed here shows the extraordinary versatility of fossil fuel-generated air pollutants, climate change, and petrochemicals in harming the health of children. Each by itself is capable of causing lasting damage to young developing brains and bodies. Through their combined and cumulative effects, they inflict an even greater toll on children's health and future well-being—greater than any other threats today. Children from all constituent and economic strata are being affected. Knowing what we know now about the scope of the existing health damage and future risks to their health, we cannot shrug our shoulders and look away. Rather, we can now envision the tremendous benefits for the health and well-being of all our children from eliminating these exposures. We then become players, no longer bystanders, in the urgent work of transitioning from fossil fuel.

CHAPTER 4

Children Are Not All Equal

INTRODUCTION

In the preceding chapter, we saw the versatility of air pollution and climate change in inflicting harm on children's brains and bodies. All children are at risk, but a disproportionately heavy burden falls on poor children in developing countries and socially disadvantaged children in the richer countries; these are most vulnerable. The result is widening inequality which has created a public health crisis and destabilized society within countries and globally. No country is spared these impacts.

A pause to define key terms: in the United States "socially disadvantaged" individuals are those who have been subjected to racial or ethnic prejudice or cultural bias within American society because of their identities as members of groups and without regard to their individual qualities. Individuals who are "economically disadvantaged" are those falling below the federal poverty line, receiving government assistance, or with an unemployed principal wage earner. The same general terms apply to other countries, although the criteria differ. The term "disadvantage" is used here to encompass the array of closely interwoven social and economic factors that, along with environmental exposures, health systems and health behaviors, contribute to large health disparities globally.

Another key term is "race/ethnicity." Technically, *race* refers to physical differences that groups and cultures consider socially significant. For example, people might self-identify as African American or Black, Asian, European American or White, Native American, or some other race. Or they may self-identify as Latino or another ethnicity based on shared cultural characteristics such as language, ancestry, practices, and beliefs. Because

Children's Health and the Peril of Climate Change. Frederica Perera, Oxford University Press. © Oxford University Press 2022. DOI: 10.1093/oso/9780197588161.003.0004

these terms are often used interchangeably and participants in studies may self-identify either by race or ethnicity, we use the composite term.[1]

"Environmental injustice" is defined here as the disproportionate exposure of communities of color and low-income communities to pollution and climate change, its associated effects on health and environment, and the unequal environmental protection and environmental quality provided through laws, regulations, and governmental programs. Left out of this definition is the additional injustice that children—who are most vulnerable, depend on adults for protection, and did not create the problems—bear the brunt of air pollution and climate change.[2]

In this chapter we explore the glaring socioeconomic and racial/ethnic disparities in exposure and health impacts experienced by children. In disadvantaged populations the harm from disproportionate exposure to air pollution and impacts of climate change is magnified by susceptibility due to inadequate nutrition, lack of adequate healthcare, social support, and housing, alongside psychological stress from material hardship due to poverty, violence, and racism. Stress "gets under the skin," setting up biologic responses that aid and abet toxic air pollutants in inflicting harm and earning the name *toxic stress*. Take the major chronic stressor: worldwide, a staggering 1 billion children, almost half of the 2.2 billion children younger than 15 years of age, are poor. In the United States, the world's most prosperous country, the child poverty rate is 16%: nearly 1 in 6 children live in a family with an annual income below the federal poverty line. More than 70% of these are children of color. Black and Hispanic children are about three times as likely as White children to be living in poverty. So, it is not surprising that there are striking systemic differences in the incidence and severity of diseases between socioeconomic and racial/ethnic groups of children. These disadvantaged groups have dramatically worse health outcomes than others, evident even at birth. This pattern laid down early in life persists throughout the childhood, adolescent, and adult years.[3]

Finally, "intergenerational injustice" refers to the unfair burden borne by the younger generation and those to come. The burden will be carried through their lifetimes and by their progeny unless we, the ones responsible, take strong action now. The ethical principle of preventing harm to future generations ("our grandchildren") is widely held as a core value, increasingly recognized by legal systems around the world. In the 2015 Paris Agreement on Climate, for example, 196 nations agreed to respect and promote intergenerational equity. Intergenerational equity is not new, however: it was embodied in the constitution of the Confederation of the Six Nations of the Iroquois, which required leaders to make decisions with the "Seventh Generation to come" in mind. Since the early 1970s, international treaties have set out responsibilities to protect future generations, especially those related to environmental protection.

DISPARITIES IN AIR POLLUTION EXPOSURE AND EFFECTS

With respect to disparities in exposure to air pollution, we see a repeat of the story for lead, which has long been a poster child and rallying cry for environmental justice. While lead poisoning crosses all socioeconomic, geographic, and racial boundaries, the burden of this disease has fallen disproportionately on children in disadvantaged communities everywhere in the world. In the United States, children in low-income communities and communities of color living in public housing units built before the banning of lead paint in 1970 were most affected. Over the past 20 years, childhood lead poisoning has declined dramatically in the United States due to regulation. Nonetheless, today, lead poisoning may be affecting as many as 1.2 million US children, mostly those in low-income families and children of color. The recent discovery of children poisoned by lead in water in Flint, Michigan and elsewhere has reignited the campaign for environmental justice.

Air pollution is also an egregious example of environmental injustice. The World Health Organization (WHO) reports that 9 out of 10 people around the world breathe air containing high levels of pollutants. "Air pollution threatens us all, but the poorest and most marginalized people bear the brunt of the burden," says Dr. Tedros Adhanom Ghebreyesus, Director-General of WHO. [4]A vivid example of disparities is the strikingly different experience of two children living in New Delhi, India, whose air pollution exposure was tracked over the course of one day in December 2020. Vehicle exhaust accounted for around 20–40% of fine particle ($PM_{2.5}$) pollution in New Delhi; the rest was largely from industrial emissions and household fires. In this ingenious study, journalists at the *New York Times* partnered with ILK Labs to compare air pollution exposure of the two children: one from an upper-middle-class family, the other from a poor family. The children kept air monitors with them during the day that took continuous measurements. As the day began, 13-year-old Monu ate breakfast near the wood-burning clay stove on which his mother cooked, then biked to a free open-air school under a bridge with heavy traffic passing close by. After her breakfast in a house with air purifiers running and where breakfast was cooked in a separate kitchen, 11-year-old Aamya was driven to a private school that carefully monitored air pollution. When the air pollution gets too high, students are required to wear masks. Monitoring of the children showed that, over the day, Monu was exposed to about 4 times as much fine particulate matter as Aamya. Figure 4.1 shows the difference in pollutant concentration from children's pollution monitors (Monu's on the left and Aamya's on the right).[5]

Using a relatively conservative estimate of the years of life lost per unit increase in air pollution, the team calculated that, if such a disparity were sustained over the long term, a child in Monu's situation could have 5 fewer years of life than an upper-middle-class child like Aamya. If air pollution levels were reduced to the WHO guideline, both children would gain: Monu about 6 years, Aamya about a year of added life. These estimates are based on the best available data but necessarily incorporate a series of assumptions, so they are not certain; they do, however, graphically illustrate the injustice of air pollution.

In rich countries also, we see the same disparity. Air pollution exposure is not unique to populations of color or low-income: more than 4 in 10 (135 million) people in the United States live in counties with unhealthful levels of either ozone or particle pollution. The number has increased in recent years in large part because of warmer weather, aberrant rain patterns, and major wildfires due largely to climate change. However, people of color in the United States are 61% more likely than White people to live in counties with high levels of particle and other air pollutants. This disparity is due to many interrelated factors: racism, class bias, land costs, unfair housing market and zoning practices, and imbalances in political power that perpetuate the siting of pollution sources such as power plants, industrial

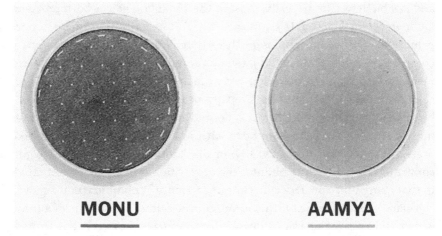

MONU   AAMYA

Figure 4.1 Photos of the small filters that were inside the children's pollution monitors.
Credit: Photo by Karan Deep Singh and Omar Adam Khan.

sources, and major roadways in or near disadvantaged communities. As a result, on average, Blacks are exposed to about 56% more PM$_{2.5}$ emissions and Hispanics to 63% more than is caused by their consumption of goods and services. Whites, on the other hand, have a "pollution advantage": they experience approximately 17% less air pollution emissions than is caused by their consumption. Specifically looking at emissions from fine particle-emitting facilities, we see that the burden in Black Americans is 50% higher compared to the overall population, with the racial disparities being even more pronounced than for poverty status. A study of more than 150,000 children in the United States found a consistent pattern of racial inequity in the distribution of major stationary and mobile air pollution sources, with Hispanic and Black children facing the highest potential exposure[6]

Europe displays the same pattern of disparities that we see globally and in the United States. People of lower socioeconomic status tend to live, work, and go to school in places with worse air quality. More impoverished areas—Molenbeek in Brussels, Tower Hamlets in London for example—are hot spots for air pollution. Apart from poverty, belonging to certain racial/ethnic or Indigenous groups in Europe means more exposure to pollution.[7]

Not surprisingly, decades of research have documented striking systemic disparities between socioeconomic and racial/ethnic groups in the incidence and severity of the diseases and developmental disorders associated with air pollution. The disparities are reflecting a host of social and environmental factors; air pollution is but one of the factors, but exposure to air pollution is a significant contributor—and it is by nature preventable.

Injustice starts even before birth with fetal exposures. As we saw in the previous chapter, both air pollution and heat have a hand in preterm birth and low birth weight. Globally, most of the 15 million babies born preterm are in lower income countries. In the United States, the preterm birth rate is more than 50% higher among Black women (almost 14% born preterm) compared to White women (9%). Groups at highest risk from maternal exposure to heat and air pollution are women with asthma and women of color, especially Black mothers. Globally, of the more than 20 million low-birth-weight babies, most are in low- and middle-income countries. But more than 310,000 babies are born with low birth weight in the United States, with a gaping disparity: 13% of Black infants are low-birth-weight compared to 7% of White infants. The disparities at birth are perpetuated in that many survivors face a lifetime of disability.[78]Infant mortality shows the same glaring disparity. This is yet another effect tied to air pollution, which takes its major toll in developing countries; however, the United States has the highest infant mortality rate of any developed country, with large variation between racial/ethnic groups. The rate is highest for Black infants (almost 11 per 1,000 live births), more than twice that in Whites. Of all infant deaths in 2017, about a third were a result of preterm-related causes.[9]

Long established as a trigger of asthma in children, air pollution is now considered to be a cause of the disease as well. As with the other outcomes, asthma prevalence in US children varies greatly by race/ethnicity: 7% of all children younger than 18 years have asthma—an average that conceals the fact that almost 14% of Black children, compared to 6% of White children, suffer from asthma. The rates are far higher in low-income, urban communities of color such as in Central Harlem, Washington Heights, and the South Bronx in New York City, where nearly 1 in 3 children report having asthma or asthma-like symptoms[10]

A study following almost 2,500 5- to 9-year-old children in California for 3 years showed that stress and low socioeconomic status conspire with air pollution to raise the risk of developing asthma and contribute to the disparities in rates. The children had no history of asthma or wheeze when the study began. Parents reported their level of stress and educational achievement (a proxy for socioeconomic status) at the beginning of the study. The research team estimated each child's exposure at his or her residence to air pollution from traffic. Parents alerted the researchers when their child was diagnosed by a doctor as having new-onset asthma. At the end of 3 years, the researchers found that the risk of developing asthma associated with traffic-related pollution was significantly higher for children of parents reporting high levels of personal stress. A similar interaction was seen between air pollution and low parental education.[11]

Air pollution affects children's immune defenses, increasing their vulnerability to infectious agents. In the United States, the incidence of pneumonia and other invasive pneumococcal diseases (infections caused by the bacterium *Streptococcus pneumoniae*) has traditionally been much higher in children of Alaska Native, Native American, and African American race/ethnicity compared to other groups. The COVID-19 crisis of 2020–2021 provides a stark example of racial/ethnic differences: 78% of the 121 COVID-19–associated deaths in persons younger than 21 years reported to the Centers for Disease Control (CDC) between April and the end of July 2020 were Hispanic, Black, and American Indian/Alaskan Native.[12]

As referenced earlier, air pollution has been linked to mental health disorders including attention deficit hyperactivity disorder (ADHD), autism spectrum disorder, anxiety, and depression. These disorders follow a familiar pattern along racial/ethnic lines. For example, the percent of Black children diagnosed with ADHD or a learning disability is higher than for White children in the United States. A study of 290,000 mental health–related visits to pediatric emergency departments in the United States concluded that Black children had a 52% higher rate of mental health visits compared with White children. The CDC's Youth Risk Behavior Survey found significantly higher prevalence rates of sad mood, suicidal thoughts, and suicidal attempts among Latino and Black youth compared to Whites. These disparities can be attributed in substantial part to lack of access to quality mental health services for children of color, who are more likely to receive inferior health services compared to their White peers. But we must also reserve blame for air pollution and stress due to poverty and racism—and to their toxic interaction.[13]

DISPARITIES IN EXPOSURE AND HEALTH IMPACTS OF CLIMATE CHANGE

Climate change affects the physical and mental health of children everywhere, even in the more affluent countries. Every harmful climate change impact—more frequent severe storms and floods, heatwaves, forest and wildland fires, droughts, vector-borne diseases, contaminated water and food, conflict, displacement, and forced migrations—disproportionately affects the health of marginalized and disempowered children. All these climate impacts can interact with underlying inequality due to poverty, poor health status, racism, and discrimination, such that the health of more vulnerable children is harmed "first and worst." In a downward spiral, each subsequent disaster or stressor is more damaging than the one before, as the ability to cope and recover is diminished. The inequality exists both between and within countries. The term "climate gap" has been coined to refer to the disproportionate effects of climate change on disadvantaged groups—a gap that is sure to grow in coming years unless policymakers intervene to change this course.

Floods and Severe Storms

The risk of exposure to severe storms and floods is universal but is greatest for poor and marginalized families and children who tend to live in areas that are especially vulnerable to severe flooding from major storms and who have the least capacity to withstand or recover from these events. A child living in poverty and without adequate housing, food, water, and sanitation before a crisis will not only be more affected by a flood or storm but also at even greater risk in a subsequent crisis. In the past two decades, more than

7,000 major environmental disasters caused more than 1 million deaths worldwide and trillions of dollars in damage, mostly in developing countries. The number of people who died per disaster in low-income countries was more than three times that number in high-income countries. Worldwide, almost all the estimated 67 million children directly affected by weather-related disasters every year in the past 10 years—and most of the children who died—were in lower income countries. A single event, Typhoon Haiyan in 2013, directly affected nearly 6 million children in the Philippines.[14]

Even within countries that are more affluent and experience fewer disasters, the impacts of disasters are strikingly unequal. In the United States, counties, cities, and neighborhoods with large numbers of Black and Hispanic residents have suffered disproportionately from hurricanes and floods because they are more often located in damage-prone areas and lack the resources to recover quickly from those disasters. This is the result of long-lived policies that clustered minorities in undesirable areas such as floodplains. The concentration of minorities in urban areas that are vulnerable to flooding can be traced to "redlining," a practice instituted in the 1930s that marked areas as risky real estate investments because residents were Black. Redlining segregated races and pushed Black people into dense urban housing. These neighborhoods are now liable to urban flooding because they have few parks, open spaces, or tree canopies to absorb excess water. As a result, people of color make up 81% of the population in the 10 counties deemed by the National Oceanic and Atmospheric Administration (NOAA) most vulnerable to natural disasters, although they represent only 39% of the United States population. They became even more threatened by climate disasters as the coronavirus pandemic weakened their resilience.[15]

The effects of Hurricane Katrina in Louisiana vividly illustrate climate injustice. Katrina devastated New Orleans, hitting the primarily African American neighborhoods of the East and Lower Ninth Ward hardest. Many Black and high-poverty neighborhoods were located in these low-lying, less protected areas of the city and were both more exposed and more vulnerable. Forty-six percent of the people in damaged areas were Black compared with 26% in undamaged areas; 84% of missing people were Black in a city that was 68% Black. More than 80% of the homes in New Orleans lost in Katrina belonged to African American residents. Black children were more likely to be exposed to Katrina's floodwaters, to be evacuated, and to have mental health problems. Trauma symptoms in youth were elevated 30 months later in areas of the city that were most heavily impacted by the hurricane. Many children spent years in formaldehyde-laced trailers provided by the Federal Emergency Management Agency (FEMA) that were meant to be temporary housing.[16]

Mold in homes as a result of disasters also observes the general rule of inequality. African American children in the United States are more likely to live in low-quality housing and have higher rates of sensitization to mold. Major storms cause flooding and excess moisture in homes and lead to the proliferation of mold, which is a notorious trigger of asthma attacks. In the aftermath of Katrina, the majority of the 112 homes in the New Orleans area assessed by the CDC had "visible or heavy", mold growth. Overall rates of asthma in children from the New Orleans area were increased post-Katrina, especially among children of color and underprivileged children affected by post-Katrina flooding.[17]

Europe displays the same general pattern of disparities with respect to the impacts of climate change. High-income households are responsible for the largest share (37%) of the continent's overall carbon footprint and the lowest income households the smallest (only 8%). Yet, low-income families in Europe are most exposed and most vulnerable to climate change. Apart from poverty, belonging to certain racial/ethnic or Indigenous groups means more exposure to disasters from climate change. One such group is the 10 million Roma people living in central and eastern Europe. Their communities are frequently located on or near landfills or contaminated industrial land, lack running water and sanitation in homes, and do not have access to education. Roma people have suffered especially severely during floods. In June 1998, the Slovakian Roma community of Jarovnice in the Slovak Republic experienced one of the worst floods in its history. According to one report, some 140 Roma homes were affected, compared with 25 non-Roma homes: of the 47 people who were killed, almost all were Roma. Social advantage was reflected in topography: 42 of those who died lived in a shanty town in the valley of the Mala Svinka River that flooded, while non-Roma lived in the village above the valley. Another report placed the number of dead at more than 60, overwhelmingly children: 44 children and 16 adults. As usual, children bore the brunt.[18]

The summer of 2020 was devastating for millions of children and families in Bangladesh, Bhutan, India, and Nepal who were affected by weeks of torrential monsoon rains, widespread flooding, and deadly landslides. UNICEF estimated that more than 4 million children were impacted and in urgent need of life-saving support, with many millions more at risk. These included an estimated 1.3 million children in Bangladesh, 2.4 million children in India, and 5,000 children in Nepal. Once again, almost all the families and children affected were of low socioeconomic status and most vulnerable to the loss of homes, crops, and water and sanitation infrastructure.[19]

As a result of climate change, Indigenous peoples of Europe, like the Sámi (also known as Saami, historically known in English as Lapps or Laplanders) are losing their livelihood, way of life, and culture. The Sámi people, who number about 80,000, live in Norway, Sweden, Finland, and Russia, largely gaining their livelihoods from coastal fishing, fur trapping, and sheep and reindeer herding. Currently about 10% of the Sámi are engaged in semi-nomadic reindeer herding. Snow, or *muohta*, covers the Sápmi region— home of the Sámi people—8 months a year and determines the success of the reindeer herd. There are at least 360 words for snow in the Sámi culture. One is the word *guohtun*, which describes both snow and nutrition conditions for reindeer. Conditions for *guohtun* are no longer predictable: not only does snow arrive later, but the amount varies and the snow has a different structure. In a warming climate, the unique cultural livelihood and knowledge system of this ancient people is likely to fade away.[20]

Heat

The world's lowest income countries are already experiencing greater increases in the occurrence of extreme temperatures compared to the highest income countries; this has been the case for more than two decades. Low income and socially marginalized populations everywhere are more at risk because of their limited access to climate-controlled housing and shelter from heat outdoors. In the United States, too, exposure to extreme heat is unequal along both socioeconomic and ethnic/racial lines. Many of the same disadvantaged groups most affected by storms and flooding are also vulnerable to extreme temperatures: they live in urban "heat islands" created by extensive development using concrete and pavement and that lack green spaces and trees to cool them. In the 10 counties that NOAA rates as the most vulnerable to extreme heat, the population is on average 67% non-White, far higher than the 39% of the US population they represent. A new study shows that urban neighborhoods denied municipal services and support for home ownership through redlining are now hotter, by as much as 7°C (almost 45°F), than their non-redlined neighbors in more than 90% of 108 US urban areas analyzed.[21]

For example, in the 1930s, federal officials redlined many neighborhoods in Richmond, Virginia, one of which was the mostly Black neighborhood of Gilpin. As recounted in the *New York Times*, to escape the heat of summer, Sparkle Veronica Taylor, a 40-year-old resident of Gilpin, often walks with her two young boys more than a half-hour across Richmond to a tree-lined park in a wealthier neighborhood. Her local playground lacks shade, leaving the gyms and slides to bake in the sun. "The trek is grueling in summer temperatures that regularly soar past 95 degrees, but it's worth it to find a cooler play area, she said."

"The heat gets really intense, I'm just zapped of energy by the end of the day," said Ms. Taylor, who doesn't own a car. "But once we get to that park, I'm struck by how green the space is. I feel calmer, better able to breathe. Walking through different neighborhoods, there's a stark difference between places that have lots of greenery and places that don't."[22]

Indigenous groups also bear a disproportionate share of the impacts of rising temperatures. Alaskan and Pacific Northwest Native communities that have long subsisted on traditional hunting and fishing have seen an unprecedented drop in their food supply because of the damage to fish and wildlife species inflicted by higher temperatures. Many Alaskan natives may have to relocate entire villages due to thawing permafrost caused by global warming.

Just as we saw with the effects of air pollution on birth outcomes and infant deaths, heat-related effects on these outcomes are greater in disadvantaged populations globally. A study of 53 developing countries spanning the globe found that the estimated impacts of hot days on infant mortality are 10-fold greater than estimates from rich countries. As we saw, in many cities in rich countries, people of color and lower socioeconomic status have disproportionate exposure to heat, and consequently greater health risks.[23]

Forest and Wildland Fires

Forest and wildland fires, which have become more frequent with climate change, also exact their greatest toll on the least advantaged. An estimated 7.6 million children in the United States are affected by wildfire smoke annually, many of them in the Southeast, Pacific Northwest, and California. Children with preexisting asthma are more vulnerable to the respiratory effects of the smoke. Patricio, a seventh grader, lives in a largely Latino neighborhood of California's Central Valley flanked by busy highways and agricultural fields; the region has the highest ranking in particulate matter pollution in the state. Patricio has asthma, and, even when there are no fires, he struggles to breathe when the air pollution level becomes too high. In an interview Patricio said, "If you had a child with asthma or any person in your household with asthma and you wanted to move into this area, it's not a good idea. I don't recommend it."[24]

Prompted by the vivid evidence from Hurricane Katrina of disparities in impacts, researchers at the University of Washington and the Nature

Conservancy created an index that characterized a community's ability to adapt to wildfires; they applied it to more than 70,000 census tracts across the United States. They concluded—no surprise—that wildfire vulnerability is spread unequally across racial and ethnic groups. Twenty-nine million Americans live with significant potential for extreme wildfires, a majority of whom are White and socioeconomically secure; however, for the 12 million socially vulnerable Americans, a wildfire event could be devastating. In their analysis, census tracts that are majority Black, Hispanic, or Native American have about 50% greater vulnerability to wildfires compared to other census tracts. In the previous chapter, we saw that wildfire smoke is linked to low birth weight, preterm birth, asthma attacks, and other respiratory illness. These risks from wildfire smoke are increased in children living in low-income communities and communities of color that have had historically high exposure to unhealthy air and high rates of underlying health conditions. In the summer of 2020, as the Western United States faced an already active wildfire season, public health officials warned that smoke exposure was likely to worsen the coronavirus pandemic by increasing susceptibility to the virus that has already disproportionately affected Black and Latino children.[25]

In addition to causing physical trauma, wildfire disasters have impaired children's mental health. High levels of stress and symptoms of posttraumatic stress disorder (PTSD) have been reported in youth affected by wildfires from California to Australia. As with other disasters, lack of adequate resources to cope with and recover from a fire increases the likelihood of mental health problems.

Drought and Malnutrition

Another striking inequality is seen in the occurrence of malnutrition in children due to drought. Most of the burden has been on people living in rural areas of developing countries. Regions of Africa, Central and South America, Afghanistan, and other parts of Asia have been hit especially hard. However, Australia has experienced severe drought in recent years, as have the Western and Southeastern regions in the United States. The usual pattern of inequality is seen: in California during the 5-year drought from 2012 to 2016, communities of color and low-income people living in tribal, rural, and farming communities bore a disproportionate share of the drought's burden. In addition to experiencing greater food insecurity, these groups are most likely to lose access to clean tap water as droughts become more common and severe.[26]

Drought due to climate change is a threat multiplier for hungry and undernourished children in several ways: it affects not only the availability and affordability of food, but also its nutritional value. Children in developing countries have been especially affected by malnutrition and its dire consequences for their health and brain development. In 2020, globally, 149 million children under the age of 5 (22%) were stunted; another 14 million suffered from severe wasting. Although malnutrition is rare in the richer countries, undernutrition is not. Stunting affects only 3.4% of US children, but 15% of US households with children are food insecure and at risk of undernutrition. The rates of food insecurity differ sharply by race/ethnicity and socioeconomic status: Black households (almost 22%), Hispanic households (17%), White households (8%). Households below the poverty threshold have the highest rate of all (28%). The coronavirus pandemic has worsened the problem as millions of Americans lost their jobs: in late June 2020, more than 27% of households with children (about 14 million children) were food insecure.[27]

Infectious Disease

Infectious vector-borne diseases to which climate change is contributing, like malaria and dengue carried by mosquitos, take a terrible toll on children in low-income countries. So do cholera and other infectious diarrheal diseases caused by crop and water contamination from heavy rainfall and floods. Diarrheal diseases account for 1 in 9 child deaths worldwide, making diarrhea the second leading cause of death among children under the age of 5. More than half a million children die from diarrhea each year. These diseases are mostly rare in the richer countries. However, the burden of diarrheal disease in the United States falls more heavily on children in Black families who lack adequate nutrition and have less access to preventive care and treatment and where Black mothers often face barriers to breastfeeding. As we learned from the tragic lead poisoning of children from drinking water in Flint, Michigan, water and sanitation infrastructure is notoriously poor in historically Black communities.[28]

Trauma and Stress from Migration and Displacement

Climate change is an increasing driver of migration, as more and more people in low-income countries are forced to flee climate-related drought, rising sea levels, severe storms, fires, and conflict. In 2019, 20 million people were displaced due to extreme weather events alone. It is predicted that the number of environmental migrants could grow up to 1 billion by 2050 if climate change is not forcibly addressed. Conflict around the world is being linked to impacts of climate change—ecological collapse, resource depletion, and temperature change—so that climate change is now being seen as a catalyst for conflict, not just a threat multiplier. In Syria, the 7-year war has displaced more than 12 million people, as many as 6 million of them children. Many children have died. The tragic image of a 3-year-old Syrian child, Alan Kurdi, lying on a Turkish beach captured the world's attention in 2015 (Figure 4.2). The child had drowned in a failed attempt by a group of refugees to sail to the Greek island of Kos.[29]

Such conflicts have not only directly exposed children to violence and trauma, but also have torn families apart, interrupted schooling, cut off access to healthcare or food, and eliminated the jobs that families depended on for a living. Children caught in and fleeing zones of conflict have suffered physical injuries, psychological damage, developmental delays, malnutrition, and lost years of education. The effects often persist into adulthood and limit the person's ability to make a living; they may also affect the next generation and beyond. In addition to physical trauma, displacement and

Figure 4.2 A Turkish police officer stands next to the body of a young Syrian boy.
Credit: Nilufer Demir/Dogan News Agency/AFP/Getty Images.

conflict have driven sexual violence and sexual trafficking, especially affecting poor and marginalized women and children.

Displacement and forced migration place disadvantaged children and adolescents at high risk of chronic stress, anxiety, depression, and posttraumatic stress. The effects have been felt mainly in developing countries and among the least advantaged children within them. Although rich countries have not experienced major climate-related internal migrations, unless unchecked, heat and extreme weather events may force many millions of Americans to migrate north in the next decades. Children in poverty and children of color will be most affected.

Although not itself greatly affected by internal migration due to climate change, Europe has been shaken by the huge numbers of migrants entering the continent, more than half of them climate refugees. In 2015, the number of people who applied for asylum in Europe surged to a record 1.3 million. Many had fled war-torn countries such as Syria, Afghanistan, or Iraq that experienced a change in climatic and weather conditions. The number has been dropping since, but, in 2019, more than 123,000 migrants and asylum seekers arrived in Europe, mostly by sea. As many as a third were children who are vulnerable to abuse and other grave forms of violence during and after their journeys. For many of those who survived the perilous journey to Europe, a life of poverty and deprivation awaited them. Most of the 3 million Syrians now living in Turkey are hosted in communities that are themselves poverty-stricken. Almost 1.5 million are children, hundreds of thousands of them out of school. There are now more than 72,000 refugees and migrants stranded in Greece, Cyprus, and the Balkans, including more than 22,500 children. The children are increasingly showing signs of deep psychological trauma as a result of the distress and deprivation they have experienced during and after their journeys. These children start the journey in poverty and suffering and often meet the same fate in Europe.[30]

PETROCHEMICALS FROM FOSSIL FUEL: ANOTHER SOURCE OF DISPARITIES

Like air pollution and climate change, petrochemicals are disproportionately affecting children in marginalized communities in the United States and in other countries. Contamination of air, drinking water, soil, food—and bodies—by petrochemical-derived toxicants is ubiquitous, as evidenced by biomonitoring studies in adult blood, umbilical cord blood, and urine taken from people around the world. Globally, toxic chemical pollution is a major threat to children's health, disproportionately impacting the most disadvantaged. There are few biomonitoring data in developing countries; however, studies in the United States measuring petrochemical-derived toxicants in blood or urine found higher levels in pregnant women and children of color and low socioeconomic status. A key reason is that the chemicals have historically been produced in those communities, which have suffered the heaviest concentrations of industrial pollution as a result. In fact, the environmental justice movement arose out of early observations of this phenomenon. For example, the multitude of petrochemical plants located along an 85-mile stretch of the Mississippi River in Louisiana, labeled "Cancer Alley," were built in largely African American communities. The 125 companies in this region manufacture a quarter of all petrochemical products made in North America. The parish of St. James, Louisiana, plunk in the middle of "Cancer Alley," has 32 petrochemical plants—one for every 656 residents. It has had the distinction of being the most polluted parish in Louisiana. These plants release large quantities of toxic substances into the atmosphere and generate waste that has contaminated the water. According to the US Environmental Protection Agency (EPA), tens of thousands of people living beside petrochemical plants along the Mississippi River are exposed to toxic chemicals at rates that are among the highest in the United States. Louisiana has ranked No. 2 in toxic emissions, just behind Texas, almost every year since 1988, when the EPA began requiring industry to tally its pollution. The community group Rise St. James is fighting to block two huge new petrochemical plants in their community.[31]

Taking our four model chemicals for illustration, biomonitoring studies in blood and urine have demonstrated widespread exposure to chlorpyrifos, phthalates, polybrominated diphenyl ether (PBDE) flame retardants, and bisphenol A (BPA) in the United States and globally. In the United States, measured levels of pesticides and PBDEs are higher in non-White racial/ethnic groups and populations with lower household income and educational attainment, including among pre-adolescent girls, adolescent girls, and pregnant women. In 2014–2016 PBDE levels in blood and placental tissue were highest in Blacks, followed by Hispanic/Latina, than Whites and Asians. Levels have come down since the phase-out of certain PBDEs, but because of their persistence these "legacy chemicals" will likely remain ubiquitous in the population. Similarly, among reproductive-aged women and pregnant women, Blacks and Hispanic/Latina women have higher metabolite concentrations of common phthalates found in personal care products than do White women. Following the same pattern, higher levels of BPA have been measured in African Americans and people with lower incomes than in the population at large.[32]

These exposure disparities result not only from living near production facilities but also from greater exposure to the chemicals during daily living, through food, consumer products, and personal care products. For example, in low-income communities and communities of color, the higher prices and lack of availability of organic foods mean that there is greater dietary exposure to neurotoxic pesticides like chlorpyrifos. There is greater use of cosmetics and personal care products containing phthalates due to the higher cost of alternative products and lack of awareness of phthalate toxicity. While more affluent people have been able to replace older couches and other furniture containing banned PBDEs, low-income families cannot afford replacements and have often been forced to buy second-hand furniture that contains those flame retardants. Similarly, although BPA was phased out in the mid-2000s, BPA-containing food containers and other products continue to be used by people who are less aware of the toxicity and less able to afford substitutes.

Mental Health

This topic deserves special note as we increasingly recognize that air pollution, the impacts of climate change, and toxic petrochemicals can exert cumulative effects that are felt most by children in disadvantaged communities. We saw in the preceding chapter that these exposures can combine to increase the risk of common mental health disorders in the young: anxiety, depression, ADHD, and PTSD.

These exposures have created widening disparities in mental health along racial/ethnic and economic lines. Not only are the exposures higher and more common in disadvantaged communities, but they also are acting in conjunction with underlying vulnerability due to poverty and discrimination. Their disproportionate and accumulated impacts are mirrored by the higher rates of conditions such as ADHD and mood disorders in US children of color and low-income status. Similarly, although "climate anxiety" is felt by children of all socioeconomic and ethnic groups, children in communities that lack the financial and social resources to prepare for, manage, and recover from climate disasters are more vulnerable to mental health disorders.

WHAT EXPLAINS SUCH PERVASIVE DISPARITIES?

What is the biological basis for the striking differences in health outcomes between racial/ethnic and socioeconomic groups? Genetic differences do not explain them. Unlike sickle cell anemia, for most health conditions genetic variation between racial groups does not correlate with differences in their prevalence. Instead, disparities can be ascribed to a complex of social and environmental factors such as poverty, racism, unequal access to healthcare, lack of education, pollution, and the impacts of climate change. These social and environmental exposures are concentrated in racial/ethnic minorities and those who live in socioeconomically disadvantaged circumstances. Adverse social conditions lead to stress, which has been identified as one of the major determinants of health disparities. We saw how social stress is not only harmful by itself but also increases vulnerability to toxic environmental exposures.

How does stress get "under the skin"? As explained by a leading expert in the field, the late Professor of Neuroscience at Rockefeller University Bruce McEwan, not all stress is bad for you. Good stress comes from rising to a challenge, causing you to feel exhilarated when your body and brain are working properly to help you do so. Toxic stress comes from not having the internal and external resources to cope with unpredictable, threatening situations or more minor daily demands ("hassles"). The result: inflammation and suppression of immune function.[33]

To understand why, we look to our evolutionary heritage. In *Homo sapiens* and other mammals, the "fight-or-flight" response evolved as a survival mechanism in life-threatening situations. Unfortunately, this carefully orchestrated response—useful when one is confronted by a saber-toothed tiger—can cause an individual to overreact to environmental and psychological stressors that are not life-threatening. You recall that the nervous system is composed of the *central nervous system* (including the brain and spinal cord) and the *peripheral nervous system* that connects the central nervous system to the limbs and organs. A key part of the peripheral nervous system is the *autonomic nervous system* that operates largely unconsciously to regulate bodily functions through the countering action of its two branches. One branch is the *sympathetic nervous system,* colloquially known as the "quick response, fight-or-flight" system: it acts like an accelerator to increase energy to the large muscles and raise heart rate, respiratory rate, and blood pressure in response to a challenge. The other branch, the *parasympathetic nervous system*, is the more slowly activated "calming" system that acts like a brake after the danger has passed.[34]

In the acute response to a stressful event, an area of the brain (the amydala) sends a distress signal to the so-called "command center" (the hypothalamus) that in turn sends signals to the adrenal glands. These glands respond by pumping the hormone epinephrine (adrenaline) into the bloodstream, increasing energy to the large muscles and raises heart and respiratory rates. Normal immune function is temporarily shut down, increasing inflammation. As the initial surge of adrenaline subsides, if the brain continues to perceive danger, a second hormone is produced by the adrenal glands to keep the body on high alert. This is cortisol, the so-called "stress protein". The body thus stays revved up. When the threat passes, cortisol levels fall, causing the parasympathetic nervous system to put on the brakes, calming the stress response.[35]

Normally, through its carefully orchestrated fluctuations, cortisol plays a central role in keeping the inflammatory and immune responses in proper check. But chronic stress leads to persistent overproduction of cortisol by the adrenal glands, resulting in decreased sensitivity of tissues to the regulatory effect of the hormone and triggering a cascading inflammatory response. In its turn, inflammation produces a highly reactive form of oxygen that, if not neutralized by antioxidants, can cause damage to major organs such as the liver, heart, and brain, as well as to the body's hormone, immune, and cardiovascular systems, ultimately leading to disease. Thus, chronic or repeated exposure to stress leads to wear and tear on the body, increasing the likelihood of disease and damage to developing brains.

We can now readily appreciate that stress due to poverty or racism and toxic environmental exposures can collaborate to increase health disparities. The timing of exposure is critical, with pregnancy being an especially vulnerable time for both mother and child. For example, psychosocial stress and air pollution not only independently increase risk of preterm birth, but they also interact to amplify the harm to certain groups. An analysis of birth certificate data for more than 37,000 deliveries in California found that air pollution was associated with preterm birth, largely among women in the lowest socioeconomic stratum. The interactive effects of these early exposures can be seen on brain development of the children as well. In our study of New York City mothers and their children, the negative effect of prenatal exposure to PAH air pollutants on IQ scores at age 7 was significant mainly among children whose mothers had reported experiencing hardship in the form of unmet needs for food, housing and/or clothing during pregnancy or during the child's early years. Our parallel study in Poland, in a cohort of women and children of higher socioeconomic status than the New York City cohort, found that the children of mothers who reported psychological distress in response to stressful circumstances during pregnancy and were also exposed to higher levels of PAHs had a greater likelihood of anxiety and depression symptoms at 9 years of age.[36]

It is now possible with advanced imaging techniques to visualize the harmful effects of poverty, stress, and adverse childhood events on brain development and function. Studies using magnetic resonance imaging

(MRI) of infant's and children's brains to examine the effect of prenatal exposure to maternal stress on offspring brain development have observed brain changes from birth all the way to adulthood. Offspring of more highly stressed women had changes in specific brain regions that play an important role in making decisions, regulating social behavior, processing sensory input, and processing memories. Other parts of the brain responsible for motor skills, learning and memory, and the fight-or-flight response are also affected. The signal changes observed in these imaging studies have corresponded to problems in behavioral, cognitive, and emotional development, as well as susceptibility to psychiatric disorders.[37]

In one such study, the levels of the stress hormone cortisol were measured in hair from 78 women sampled shortly after delivery. The women's infants underwent a series of brain scans while they slept. Maternal cortisol affected babies in different ways based on their sex: boys showed alterations in the structure of their amygdala (the tiny almond-shaped components of the brain involved in processing fearful and threatening stimuli). Girls, on the other hand, displayed changes in the way the amygdala connected to other nerve networks. Given the fundamental role of the amygdala in emotional regulation, these changes could help explain why children whose mothers experienced high levels of stress during pregnancy may be more likely to have emotional issues in later life.[38]

Another group of researchers used MRI to investigate the effect of poverty during childhood on brain development in 145 children aged 6–12 years who had been followed since preschool. Exposure to poverty during early childhood was associated with smaller volumes of cortical gray matter (containing the nerve cell bodies), white matter (connecting the brain cells), hippocampus, and amygdala in the children's brains. These findings provide a "brain basis" for the observations that children exposed to poverty tend to have poorer cognitive outcomes and school performance as well as higher risk for antisocial behaviors and mental disorders.[39]

Research has also shown that adverse childhood experiences can alter the architecture of the developing brain. In a 15-year cohort study, about 200 children were screened from day care centers and preschools in the St. Louis, Missouri, metropolitan area and underwent repeat brain imaging at school age through adolescence. Once again, the amygdala was one of the targets: preschool adverse experiences were significantly associated with reduced volumes of this small but powerful structure that plays a primary role in emotional development and response to stress. Another MRI-based study—this time of 119 African American low-income young adults from rural Georgia—showed that the effects of these adverse experiences can persist into adulthood. Their severity was associated with reduced volume of subregions of the amygdala; this was linked, in turn, with increased symptoms of anxiety and depression in the young men and women.[40]

Fortunately, experimental and human studies have shown that the brain is plastic and able to reverse structural changes after termination of stress. Interventions to enrich the environment have been found to reverse effects of early life adversity. [41]

Taken together, the scientific advances of the past decades have given us deep insights into the causes and biological basis of the unequal burden of illness and impairment borne by children who are socially or economically disadvantaged due to circumstances beyond their control.

CONCLUSION

In reading this chapter, we may have been shocked at the degree to which the health burden from fossil fuel is unequally borne by children who are most vulnerable. Study after study has shown us that children in developing countries and children in disadvantaged communities in the richer countries are hurt first and worst. Putting all this information together, we see that policies to wean ourselves from fossil fuel must be holistic, addressing both the disparities in social conditions and exposures to air pollution, toxic chemicals, and impacts of climate change. The goal must be to prevent their cumulative effects on children's brains and bodies, beginning before they are born.

In the next chapter, "Power and Voice," we will see how the environmental justice and climate justice movements in the United States and worldwide have emerged in recognition of disparities in exposure and impacts. They have been gaining power steadily over the past several decades and now play a pivotal role in shaping climate policies and judging their worth.

PART II

What We Can Do About It

CHAPTER 5

Power and Voice

INTRODUCTION

It is remarkable how the many voices in this chapter, coming as they do from different sectors of society and from such different life experiences, are sounding the same urgent call for action on climate change and the vital need to address injustice across racial/ethnic and socioeconomic groups and generations. All are realistic about the magnitude of the problem but hopeful that action to avoid catastrophe is possible. Human health and its connection to the health of the planet have become the focus, replacing that old symbol of climate change—a polar bear standing on a melting iceberg. All call up the same scientific consensus on the human causes and extent of climate change. But they recognize that facts are not enough and that their messages must reach people on an emotional level. We will see how the environmental justice and Black Lives Matter (BLM) movements have transformed our thinking about climate change and climate justice. We will hear from Indigenous activists, religious leaders, scientific experts, environmental and climate activists, and artists. We will see the growing intensity and power of youth who have so much at stake if adults do not act now.

THE ENVIRONMENTAL JUSTICE AND CLIMATE JUSTICE MOVEMENTS

These two pivotal movements emerged sequentially in the past century in response to the growing awareness of the egregious disparities

Children's Health and the Peril of Climate Change. Frederica Perera, Oxford University Press. © Oxford University Press 2022. DOI: 10.1093/oso/9780197588161.003.0005

in environmental exposures and health between racial/ethnic and socioeconomic groups, as described in the preceding chapter. They have been gaining power steadily over the past several decades, with a boost during the time of the COVID-19 pandemic as the public became aware that people in disadvantaged communities were being disproportionately affected by the illness. The movements now play a central role in shaping environmental and climate policies and judging their effectiveness. Both place children at the center of their arguments for action.

The environmental justice and climate justice movements have their roots in the theory of *social justice*—the concept that a society or an institution should be based on the principles of equality and solidarity, the protection of human rights, and the recognition of the dignity of every human being. Originally proposed in 1840 as a religious concept by the Jesuit scholar Luigi Taparelli, social justice was later incorporated into the teachings of the Catholic and Protestant churches. In 1971, the political philosopher at Harvard, John Rawls, elaborated the theory, defining the principle of justice as "fairness." Thus, the definition of environmental justice is "the fair treatment and meaningful involvement of all people regardless of race, color, sex, national origin, or income with respect to the development, implementation and enforcement of environmental laws, regulations, and policies." Begun as an American concept in the latter twentieth century to deal with the phenomenon of differential siting of polluting sources in communities of color and low-income communities, it was less a philosophy than a call to action, modeled on the civil rights movement of the 1960s.[1]

In 1991, sociologist Robert Bullard was instrumental in organizing the Leadership Summit at which the Principles of Environmental Justice were adopted. (He is often called the father of the environmental justice movement.) The Principles stated, somewhat lengthily,

We, the people of color, gathered together at this multinational People of Color Environmental Leadership Summit to begin to build a national and international movement of all peoples of color to fight the destruction and taking of our lands and communities, do hereby re-establish our spiritual interdependence to the sacredness of our Mother Earth; to respect and celebrate each of our cultures, languages and beliefs about the natural world and our roles in healing ourselves; to insure environmental justice; to promote economic alternatives which would contribute to the development of environmentally sage livelihoods; and to secure our political, economic and cultural liberation that has been denied for over 500 years of colonization and oppression, resulting in the poisoning of our communities and land and the genocide of our peoples, do affirm and adopt these Principles of Environmental Justice.[2]

The linked principles of sustainability and intergenerational equity are central to the environmental and climate justice movements. In 1983, recognizing the extensive deterioration of the human environment and natural resources and to rally countries to pursue sustainable development together, the United Nations established the Brundtland Commission. It was named after the Chairperson, Gro Harlem Brundtland, the former Prime Minister of Norway who had a strong background in the sciences and public health. The report, "Our Common Future," published in 1987, defined sustainable development as "development which meets the needs of current generations without compromising the ability of future generations to meet their own needs." The report was a response to the conflict between globalization of economic growth and the growing evidence of global ecological damage. It contained prescriptions for long-term environmental strategies to achieve sustainable development that met the essential needs of the world's poorest people while ensuring intergenerational equity. The report called for a unification of the goals of economic growth, environmental protection, and social equality. Years earlier, the National Environmental Policy Act of 1969 had committed the United States to sustainability, declaring it a national policy "to create and maintain conditions under which humans and nature can exist in productive harmony, that permit fulfilling the social, economic and other requirements of present and future generations." However, intragenerational equity was not new, having been embodied in the constitution of the Confederation of the Six Nations of the Iroquois, which required leaders to make decisions with the "Seventh Generation to come" in mind. More recently, especially since the early 1970s, international treaties have set out responsibilities to protect future generations, especially related to environmental protection.[3]

VOICES OF ENVIRONMENTAL JUSTICE AND SOCIAL JUSTICE ADVOCATES

The voices of environmental justice organizations such as the Black Environmental Justice Network, West Harlem Environmental Action (WE ACT), the Climate Justice Alliance, the Environmental Justice and Climate Change Initiative, and the European Environmental Bureau are growing louder and more urgent. Their common message: solutions must be developed and implemented now to ensure that the benefits accrue to all but favor the low-income communities and communities of color that have borne the brunt of pollution and climate change. Furthermore, that they protect the rights of children and successive generations to a healthy life and a sustainable future. Their impact has been huge, resetting the debate on the future of the planet and its resources from an original focus on conservation to securing health and equity for people now living and for future generations.

Black Environmental Justice Network and Black Lives Matter in the United States

In 1978, soon after they moved to Houston, Linda McKeever Bullard, a lawyer and the wife of Robert Bullard, began filing a class-action suit on behalf of the residents of a black middle-class neighborhood who were fighting a city-approved waste landfill. This was the first time anyone had sued for environmental discrimination under the 1964 Civil Rights Act. She asked her husband, who was just out of graduate school and was teaching sociology at Texas Southern University, for his help finding the documents to support the case. They found the evidence: all five public landfills in Houston were in Black neighborhoods, six of the city's eight incinerators were in Black neighborhoods, and more than 80% of all solid waste dumped in Houston from the 1930s to 1978 was dumped in Black neighborhoods. Yet Black people then made up only 25% of the population. Seven years later, the case was dismissed. The Bullards kept at it, though, for more than four decades, fighting eco-racism and environmental injustice. Seeing how climate change–driven disasters were hammering black and brown communities, in 1991 Bullard, who is currently a Distinguished Professor of Urban Planning and Environmental Policy at Texas Southern University, co-founded and recently relaunched the Black Environmental Justice Network, a national coalition of black environmental justice groups and grassroots activists "so that our communities are not left out when it comes to climate change." Bullard also co-founded the historically Black College and University Climate Change Consortium in 2011, to mobilize student groups and leaders to help defend marginalized communities, especially in the southern United States where the vast majority of historically Black colleges and universities are located and where more billion-dollar disasters occur than in the rest of the country combined.[4]

The past 10 years have seen a dramatic increase in grassroots black activism, and none with more impact than the BLM movement. Started in 2013, as a response to police brutality and racially motivated violence against black people in the United States, it is now a global network of more than 40 chapters advocating to protect Black communities against police violence as well as for other policy changes related to black liberation, including health equity. Following the killing of George Floyd by a Minneapolis police officer, an estimated 15–26 million people participated in the 2020 BLM protests in the United States, and protests were held on three continents. Due to the movement, climate activists around the world—from small grassroots groups to large environmental ("big green") organizations—are increasingly focused on racial justice. The popularity of BLM has rapidly shifted over time. Whereas public opinion on BLM was net negative in 2018, in 2020, 55% of adult Americans and 87% of Black adults expressed support.[5]

Bullard explains that he came out of the civil rights movement of the 1960s, from Jim Crow Alabama, and has witnessed a great change, "We had marches. But today the diversity and numbers of people—especially young people—there on the marches are unprecedented. . . . There is an awakening unlike any that I've seen on this earth in over 70 years." And he is galvanized by the fact that younger generations understand that their futures depend on how we address these issues. He sees the BLM protests and climate activism as part and parcel of the fight for environmental and social justice. "It is important that when we talk about climate change, it is not just presented as parts per million, but the issue of equity should be given equal weight." He believes that it is important to humanize the climate change issue, to show that there are populations who, at this moment, are getting hurt because of climate change. Such a "real scenario" is the recent resettlement of a small Native American tribe in coastal Louisiana that lost nearly all its land to rising seas and its houses to hurricanes.[6]

West Harlem Environmental Action and the Power of University Partnerships

WE ACT was initially founded in 1988, to mobilize community opposition to New York City (NYC)'s operation of the North River Sewage Treatment Plant in West Harlem (originally planned for construction in a predominantly White and upper-middle-class neighborhood) as well as the siting of the sixth bus depot in Northern Manhattan. WE ACT has since broadened its focus to include climate justice, clean air, access to good jobs, and sustainable land use. Peggy Shepard, WE ACT's co-founder and executive director, began her career as a journalist (the first African American reporter at the *Indianapolis News*) and speechwriter. In the late 1980s, she was elected the

Democratic Assembly District Leader for West Harlem. She is currently co-chair of the first White House Environmental Justice Advisory Committee. Shepard has successfully combined grassroots organizing, environmental advocacy, and experience in community-based research to become a national leader in advancing environmental justice in urban communities.

In 1998, as my colleagues and I were establishing the Columbia Center for Children's Environmental Health, I asked Peggy Shepard if WE ACT would become our lead community partner. She agreed. Our research on the health impacts of air pollution from motor vehicles, power plants, and other fossil fuel combustion could well have disappeared into the black hole where publications in prestigious journals go to be forgotten, without benefit to the communities that are affected. But Shepard and WE ACT made sure that the data reached the ears and minds of policymakers in NYC and nationally, so that action followed.

An example is the cleanup of the diesel bus fleet—which was one of NYC's major air pollution sources—by the Metropolitan Transportation Authority (MTA). Shepard explained in a 2020 interview that, 18 years earlier, every bus depot but one in Manhattan was located Uptown and they were using the worst diesel fuel at that time. That started WE ACT on a campaign against the MTA: "We were able to use the Columbia data on the impact of the air pollution on pregnant women and their children to really hammer the MTA—and so now every bus is a hybrid and they're now beginning to showcase their first electric buses as well. It just goes to show that working on a community issue—like the fact that out of the seven bus depots in Manhattan six are Uptown—just working on that one issue resulted in cleaning up every single bus for every New Yorker citywide." Later, in 2012, WE ACT was a plaintiff along with other advocacy groups and 11 states in a lawsuit challenging the US Environmental Protection Agency (EPA) to develop more protective standards for fine particulate matter. In their deposition for the legal filing, WE ACT referenced our Center's research. Then after the EPA published a proposed rule, WE ACT testified at a public hearing in Philadelphia, again citing the Center's research to advocate for a stronger standard for fine particulate matter. The EPA Administrator signed the final rule in December of 2012, adopting the stronger standard.

In the 2020 interview, Peggy turned to the disproportionate impact of the warming climate on communities of color, noting that between 2000 and 2012 nearly half of NYC's heat-related deaths were among African Americans, even though they comprised less than 25% of the city's population at that time. Now, she added, the threat of COVID-19 meant people were being asked to stay at home in the middle of summer heat. So WE

ACT reached out to the Mayor's office and the state to advocate for more funding to the Low Income Home Energy Assistance Program to increase the use of cooling technologies in NYC homes, including through a subsidy to finance cooling-related energy bills for low-income residents. "It doesn't help having an air conditioner if you don't have the money to run it," she said.[7]

INDIGENOUS ACTIVISM

Indigenous people have always been at the forefront of fossil fuel resistance to protect their land and culture. In *This Changes Everything: Capitalism vs. Climate*, Naomi Klein writes that "Indigenous rights . . . may now represent the most powerful barriers protecting us all from climate chaos." Tribal leaders have been instrumental in the fight against fossil fuel extraction and burning—from fracking for oil or gas, to coal mining, construction of new coal plants, and laying of oil pipelines. She recounts how, in the United States, the Navajo's Black Mesa Water Coalition managed to shut down a coal-fired power plant and reduce coal mining on their land. In a territory of the Niger Delta, Nigeria, Indigenous people successfully halted the extraction of oil. Then, in opposing Enbridge's proposed Northern Gateway Pipelines, a large coalition of First Nations in British Columbia, Canada, signed the "Save the Fraser (River) Declaration" in opposition to the project. In one of the hearings, Jess Housty, a young woman who was an elected member of the Heiltsuk Tribal Council in British Columbia, testified, "When my children are born, I want them to be born into a world where hope and transformation are possible. I want them to be born into a world where stories have power. . . . To practice the customs and understand the identity that has made our people strong for hundreds of generation." The pipeline project was eventually defeated.[8]

Among the many leaders across differing African tribes, Hindou Oumarou Ibrahim, a member of the Mbororo people in Chad, stands out. The Mbororo practice seasonal movement with their cattle, spending the dry season near Lake Chad; thus, they are intimately aware of the effect of climate change on their livelihood, culture, and very survival. They depend on traditional practices and weather forecasting developed over the centuries. Ibrahim recounts how Lake Chad is a water source for 40 million people but has lost 90% of its surface area in just 40 years, with resultant forced migration of men to the cities in search of work and in response to conflicts along its shoreline. In a TED talk, Ibrahim explains that Indigenous people have thousands of years of knowledge about survival and resource management compared to Western science—which is only a few hundred years old. If we put together all the knowledge systems—science and Indigenous culture, especially knowledge held by women—she explains, we can manage resources for the long term and have a healthier planet. In an interview with *Time Magazine* in 2019 she said,

> The wisdom we hold is based on living in harmony with nature. We know how to keep the balance of nature. . . . Climate policy at the international level still does not provide enough representation for Indigenous communities. Most climate policies and solutions are not being designed for the people, by the people, with the people. Often, they are designed by experts with a master's degree or PhD. But that's not enough. A lot of expertise is local. Locals are experts of their own land in ways that academics are not. Our solutions are not only about writing reports that barely anyone in my community can make sense of. There is a misconception that if you don't know the technical words, you do not understand climate. But that's not true.[9]

Ibrahim has advocated for Indigenous communities at the local, regional, and international levels and for the integration of Indigenous knowledge with Western science to create a healthier planet. She has won numerous honors, including the Pritzker Emerging Environmental Genius Award. In the run-up to the Paris Climate talks, she served as co-chair of the International Indigenous Peoples Forum on Climate Change and then worked with the government of Chad on a plan to cut greenhouse gas emissions that incorporated traditional knowledge.

On the North American continent, Faith Spotted Eagle is a member of the Yankton Sioux Nation in South Dakota; she is a tribal leader, activist, and politician. She has devoted herself to restoring endangered and lost cultural practices of the Lakota, Nakota, and Dakota peoples; protecting Indigenous lands, and fighting tar sands development. Spotted Eagle is from a village in South Dakota, where the tribal community of White Swan lived before the United States government built the Fort Randall Dam as part of the Flood Control Act of 1944, submerging the village under 140 feet of water and scattering its residents. She still lives in the area at nearby Lake Andes, South Dakota. Spotted Eagle remembers when she was a young girl, maybe 8 years old, fishing with her father on the banks of the Missouri River in South Dakota. "My dad looked at me, and he said, 'You know, my girl . . . someday you're going to have to do something about all of this,'" Spotted Eagle said, recalling a far-off look in his eyes. "I remember sitting on that bank on that summer day and thinking, 'What am *I* going to do? I'm only 8 years old.' And he said, 'You'll see.'"[10]

Spotted Eagle would go on to become a prominent activist success-fully opposing the construction of the Keystone XL and Dakota Access oil pipelines through tribal areas. In 1994, she co-founded the Brave Heart Society, which is dedicated to restoring the culture and protecting sacred land, water, and animals against tar sands pollution. Supervised by a group of community grandmothers like herself, called the Unci Circle (Unci, pro-nounced "oonchee," means Grandmother), the Society has fought fiercely against the pipelines proposed to transport the tar sands oil from the Canadian province of Alberta to the rest of North America. This is one of the dirtiest and most dangerous oils on the planet, described as a "carbon bomb." NASA scientist James Hansen has explained that draining these vast fields would mean "game over" for the climate system. Not only have the tar sands' operations encroached on Indigenous lands in Canada, contaminating the environment and endangering the wildlife these communities depend on for their culture and way of life, but Indigenous lands are affected by the many existing and planned pipelines that would transport the tar sands oil across North America from Canada, largely for export to other countries. In 2016, Spotted Eagle became the first Native American to win an electoral college vote for president. Referring to Donald Trump's election as president, she remarked that the battle against big oil was not new: "The battle that we're fighting is 500 years old. It's about dis-possession, it's about occupying our land by a foreign country, or foreign individuals. The resistance has always been in my blood and my spirit since I was born."[10]

In 2017, she described her worldview: "It was about community. The way we're raised is that when you introduce yourself, you introduce your nation first: where you're from, where you live, your family, and lastly, yourself. We are also place-based societies. When we're indigenous to a place, a lot of our knowledge is intricate and long-standing. We know these lands, and that gives us, I believe, moral authority in a pre-colonial sense." She described the protest against the Dakota Access Pipeline at Standing Rock in 2016 as a unifying call for all tribal people to come together as nations and talk to each other. "The pipeline threatened our lifeline, which is the Missouri River. We're a community that takes water from that river, so if that water is polluted, it will be like what happened with Flint, Michigan [where water was contaminated by lead]. It's a human rights issue. Standing Rock was an opportunity for 12,000 people to stand up and say, 'We don't want this development.'" She added, "I think our alliances that we've created with non-Native people need to grow because when they heal, we heal. . . . [W]e can stand together as partners in defending the earth that we live on—it's all we have." A mother to two children and a grandmother, Spotted Eagle and the other elder women are strongly focused on passing down positive teachings and action to the next generations. They want young people to know that they are supported in taking the risks of fighting environmental pollution and climate change and to be inspired by their own courageous actions (Figure 5.1).[11]

Tara Houska-Zhaabowekwe is a member of the Couchiching First nation in Northern Minnesota, an attorney, and an environmental and Indigenous rights advocate. She describes a protest to block the entrance of Enbridge's US tar sands terminal for a pipeline that would carry more than 700,000 barrels of bitumen sludge per day through her people's territory—right through their wild rice beds—on its way to the shore of Lake Superior. The protestors chanted their message of keeping fossil fuels in the ground as the sound of the sirens grew. She writes: "Our steady resistance forms cracks in the world of profit margins. It transitions us away from self-destruction.

Figure 5.1 The activist and political leader, Faith Spotted Eagle is a member of the Yankton Sioux Nation in South Dakota and co-founder of the Brave Heart Society.
Credit: Travis Heying/Wichita Eagle/Tribune News Service/Getty Images.

We are a thorn in the side of a world that believes it must extract to exist, a bone-deep reminder there are other ways of being and people willing to take personal risk for something greater than any one individual."[12]

RELIGIOUS LEADERS

The trusted voice of spiritual leaders has been increasingly powerful in shaping public understanding and motivating governments to act on climate change. This is especially true of the communications by Pope Francis and the Dalai Lama, two charismatic figures weighing in mightily on climate change as a moral issue. Less visible, but cumulatively reaching many more people, have been the efforts of Evangelical, Islamic, and Jewish leaders, as well as those of many other faiths.

One of the most cited treatises on climate change by a spiritual leader is the June 18, 2015, Papal Encyclical, *Laudato si'* (English: Praise Be to You), subtitled "On Care for Our Common Home," released in eight languages plus the original Latin. In the Encyclical, Pope Francis sought to leverage his moral authority to draw attention to climate change as a global issue that disproportionately harms the poor. He intended it to reach everyone on the planet, not just the 1.2 billion Catholics. It was a groundbreaking document in its unequivocal adoption of the scientific consensus that changes in the climate are largely man-made, that fossil fuel has a central role in causing climate change, and that technology based on the use of highly polluting fossil fuels—especially coal, but also oil and to a lesser degree gas—needs to be progressively replaced without delay. Global capitalism based on the burning of fossil fuels has created "unsustainable consumption and mounting inequity," with the poor bearing the greatest burden from pollution in rich and poor countries alike. The injustice extends to future generations. Invoking the moral principle of intergenerational equity, Pope Francis asked, "What kind of world do we want to leave to those who come after us, to children who are now growing up?" On the day of the encyclical's release, the Pope took to Twitter to state the unvarnished truth: "The earth, our home, is beginning to look more and more like an immense pile of filth."[13]

My letter to the *New York Times*, which was published a day after the Encyclical was issued, applauded Pope Francis's framing of climate change as an urgent moral issue: "The message that global capitalism, based on burning of fossil fuel, has created shocking inequities is not new. But the persuasive call for moral engagement from such an authoritative voice has been missing. The injustice is already apparent in profound effects on the poor and especially the very young, who are most vulnerable to the vast amounts of toxic air pollutants as well as carbon-driven climate change."[14]

The power of *Laudato si'* stems in considerable part from its firm basis in the scientific evidence on climate change: the extent, causes, and solutions. The Pope and the Vatican had the benefit of consultation with leading climate experts, and it showed. A review by nine climate scientists under the Climate Feedback project concluded that the encyclical "fairly represents the present concerns raised by the scientific community." One leading expert, Hans Joachim Schellnhuber, the founding director of the Potsdam Institute for Climate Impact Research and chair of the German Advisory Council on Global Change, described the science of *Laudato si'* as "watertight" and gave the pontiff an "A" for command of the subject. Indeed, said a *New York Times* writer covering environmental science, those sections of the document could serve as a syllabus for Environmental Science 101 in just about any college classroom. I myself tested this theory in a class at Columbia, when I asked graduate students in environmental health sciences to grade the encyclical in terms of accuracy of the science: they, too, gave it an A.[15]

In contrast, US climate deniers, gave it a bad grade: "The pope ought to stay with his job, and we'll stay with ours," grumbled Senator James Inhofe, known as the grandfather of climate change deniers in the US Congress, then chairman of the Senate environment and public works committee. Rick Santorum, a devout Catholic, a conservative Republican and a former senator, told a Philadelphia radio station: "The church has gotten it wrong a few times on science, and I think we probably are better off leaving science to the scientists and focusing on what we're good at, which is theology and morality." These commenters were unaware of, or in denial of, both the scientific evidence and the fact that Pope Francis has a background in chemistry.[16]

Scientific facts were certainly important but *Laudato si'* coupled science with emotionally compelling moral arguments. In fact, prior to its release there was extensive consultation with experts in the areas of ethics, policy, and theology, as well as science. Meetings were convened at the Pontifical Academy of Sciences and the Pontifical Council for Justice and Peace. Among the experts was Schellnhuber, who said in an interview: "The hard lesson scientists have learned in recent years is that presenting the facts and data about global warming and other environmental problems has not been enough to move the public to action. The issues have become so serious that only a broad moral awakening can offer hope of solving them." Pope Francis sought to provide that moral awakening, powerfully connecting environmental science with social justice. Addressing "every person living on this planet," the Pope laid out a moral case for supporting sustainable economic growth as part of the church's mission and humanity's responsibility to protect God's creation for future generations. Stating that the deterioration of the environment and society affects the most vulnerable people on the planet, he wrote: "The countries which have benefited from a high degree of industrialization, at the cost of enormous emissions of greenhouse gases, have a greater responsibility for providing a solution to the problems they have caused. They are obliged both to reduce their own carbon emissions and to help protect poorer countries from the disasters caused or exacerbated by the excesses of industrialization."[17]

Looking back over the past 7 years, what has been the impact of *Laudato si'*? Whom did it reach, and were minds changed? The answer is mixed. There was certainly wide media coverage, including in all of the major newspapers and media outlets in the United States. This was due in large part to the Pope's popularity within and outside the Catholic Church. As a teaching document, the Encyclical has been a success, strengthening climate education in secondary schools and universities around the globe. This is particularly true in Catholic educational institutions, including the 28 Jesuit universities in the United States, which have incorporated the encyclical into their curriculum.

Globally, there was a notable "Francis effect" according to Tomás Insua, who co-founded the Global Catholic Climate Movement in 2015 that included more than 300 Catholic organizations. The movement organized about 40,000 Catholics from around the world to participate in a march demanding that world leaders take action during the Paris climate negotiations, and galvanized almost 1 million Catholics to sign a petition asking world leaders to keep global warming under 1.5°C (2.7°F) below pre-industrial levels. Insua noted that these actions were totally unprecedented because Catholics had been overwhelmingly passive on the climate issue before 2015. "What helped to connect the dots between the Catholic faith and the environment was the encyclical of Pope Francis, *Laudato si'*. . . . Mobilizing nearly 1 million Catholics for climate justice last year, that would have been absolutely impossible without the encyclical, there was no way we could have achieved anything nearly close to that."[18]

As a political move, the Encyclical was a success. The timing of its release was politically astute, coming just before the UN General Assembly's ratification of the Sustainable Development Goals on September 27, 2015, and the UN Climate Change Conference in Paris held from November 30 to December 12, 2015. *Laudato si'* itself inspired greater political cooperation leading up to these pivotal international agreements. It helped that Francis was the first pope from a non-European country. Happily, the UN

Climate Change Conference in Paris culminated in the unanimous adoption of the Paris Agreement. I was one of the speakers at the meetings held in parallel with the official Paris Climate Conference: my topic was the disproportionate impacts of climate change on the world's most vulnerable children.[19]

Despite gains, 5 years after the Paris conference, Pope Francis expressed disappointment, criticizing world governments for their "very weak" response to the climate crisis. In June 2020, he issued guidance for carrying out his climate encyclical that included calling on Catholics to divest themselves of investments in fossil fuel companies. With this new sense of urgency, in 2020, the Vatican launched a year-long commemoration of *Laudato si'* aimed at spurring global citizens to adopt more sustainable practices, and it issued a new, 7-year call to action.[20]

Laudato si' was welcomed and echoed by leaders of other faiths worldwide. Three days before the encyclical was released, the Dalai Lama wrote a Twitter message: "Since climate change and the global economy now affect us all, we have to develop a sense of the oneness of humanity." A day later, Archbishop of Canterbury Justin Welby, head of the Anglican Communion, issued a "green declaration" (also signed by the Methodist Conference as well as by representatives of the Catholic Church in England and Wales and the British Muslim, Sikh, and Jewish communities) urging a transition to a low-carbon economy and fasting and prayer for success at the December 2015 UN Climate Change Conference in Paris. That same day, the Lausanne Movement of global evangelical Christians said it was anticipating the encyclical and was grateful for it. The encyclical was also welcomed by the World Council of Churches and the Christian Reformed Church in North America. On June 28, several thousand Catholics, Protestants, Buddhists, Hindus, Jews, and Moslems marched through Rome to the Vatican to demand action on climate change and to thank Pope Francis for his encyclical.[21]

The Fourteenth Dalai Lama, Tenzin Gyatso, is the most prominent spiritual leader of the Tibetan Buddhist community. At his request, in 2011, the Mind and Life Institute convened a think tank of more than a dozen leading scientists, interdisciplinary scholars, and theologians at the Dalai Lama's private residence in Dharamsala, India. The week-long meeting was framed thus: Humanity faces an environmental crisis that imperils our very survival on earth. Yet most of us go on living as though nothing out of the ordinary is happening. What is wrong with us? What could religions do to spur us to action? And what is the particular role for Buddhists? Psychologist Daniel Goleman, PhD, the meeting moderator, framed the problem: "We're all in a kind of trance. . . . Today, we're facing a real paradox: even though we love our children as much as anybody in human history, every day, each of us unwittingly acts in ways that create a future for this planet and for our own children, and their children, that will be much worse." The result was a book entitled: *Ecology, Ethics, and Interdependence: The Dalai Lama in Conversation with Leading Thinkers on Climate Change*. In its 10 dialogues, the book laid out the scientific evidence on the consequences of human activity on the health of the planet, the ethical implications of climate change, and the nature of effective action.[22]

One of the most compelling dialogues took place between the Dalai Lama and Christian theologian Sallie McFague. The Dalai Lama began by placing his fullest confidence in scientists, calling them "gurus," a word that means spiritual teacher in Buddhism: "I think the best people to stimulate awareness about what's happening and what needs to be done are not the politicians or leaders but the scientists. They are the real gurus in these matters." McFague argued that to motivate needed action a kind of religious awakening was needed—one that rejects the culture of consumerism that is the root cause of climate change: "We need to wake up to a different worldview, one that shares all our resources with our fellow creatures.". Goleman found a middle ground, remarking that Buddhism and Christian theology, as well as philosophy and psychology, have very important perspectives to offer science: "Science documents what's happening, but it doesn't necessarily have within it the mechanisms to mobilize people to act in a skillful way."[23]

The hopeful tone of the book is striking, not only focusing on what we humans are doing wrong but also providing examples of where we are doing it right—such as in Bhutan, where 50% of the country is national park, farming is headed toward all-organic, and carbon emissions are on the decline. Nevertheless, the book ends with the Dali Lama's dire statement: "The earth is our home, and our home is on fire."

Four years later, in June 2015, after praising the Pope's Encyclical, the Dalai Lama called on fellow religious leaders to "speak out about current affairs which affect the future of mankind." He called for increased pressure on governments around the world to stop burning fossil fuels, end deforestation, and transition to renewable energy sources. He was joined by 14 of the world's most senior Buddhists leaders in a declaration titled *The Time to Act Is Now: A Buddhist Declaration on Climate Change*. It stated: "The Buddhist teaching that the overall health of the individual and society depends upon inner well-being, and not merely upon economic indicators, helps us determine the personal and social changes we must make."[24]

Like Pope Francis, the Dalai Lama has been frustrated by the slow pace of progress: 5 years later, in "Our Only Home: A Climate Appeal to the World," he declared that the survival of humanity is at stake and "simply meditating or praying for change is not enough. There has to be action." He has spoken of the children: "I'm a monk so I have no children, but people who have children have to think about how life will be for them and their grandchildren. We're at the start of the 21st century. Even now, we should be looking ahead to how things might be in the 22nd and the 23rd centuries."[25]

Like Pope Francis, the Dalai Lama is recognized throughout the world as a leading moral authority. He has met often with world leaders, travels widely giving Buddhist teachings, has appeared in many videos and films, has authored several books, and is the recipient of numerous awards, including the 1989 Nobel Peace Prize. He is sought after by the press and has a large following, well beyond the 535 million practicing Buddhists around the world. A global celebrity who has been described as a "religious rock star," the Dalai Lama consistently ranks with Pope Francis among the top 10 most popular world leaders. Climate change is but one of the issues he has advocated for: the list includes peace, the environment, economic and social justice, women's rights, and the welfare of Tibetans—all of which are closely intertwined. While he is one of the most accomplished beings on the planet, his power and attraction derive from his universal compassion and simplicity. Amid the chaos of today's world, he is undaunted and even cheerful. The American Buddhist author and academic Robert Thurman writes: "It takes a living personification of the Buddha—qualities to make our own freedom and enlightenment seem really possible. . . . This is the real meaning of the Dalai Lama's presence. It is felt by all who meet him, through whatever medium, consciously or unconsciously."[26]

Watching a recent discussion between the 85-year-old Dalai Lama and 18-year-old youth activist Greta Thunberg, I was struck by the contrast. The Dalai Lama was calmly optimistic, confident in the power of young people to push for solutions. He expressed hope for the future happiness of humanity—the oneness of seven billion human beings on the planet—but stressed that our future depends on the health of the planet, our home. In words and demeanor, Greta expressed quiet desperation, spoke of the scientific data on feedback loops in nature that cause escalation of global warming, and decried our lack of respect for nature and the environment and our failure "to think our actions have consequences, a lot of the time beyond our comprehension, things that we cannot understand, things that we cannot predict. And when these things are set in motion, we can in many cases not stop them."[27]

As to his impact on global policy, the Dalai Lama has been invited to address the UN, the US Congress, the G7 meeting in 2020, and other political bodies. Like Pope Francis, he has garnered wide respect as a moral authority. Has he had an impact on the public consciousness of the urgency of climate change? I believe so, although quantitative data are lacking. Perhaps the best indicator is that Buddhists are among the most engaged religious group on the issue of climate change.

Other world religions have not had such charismatic leaders, but their messages are similar in many respects. Islam is the second largest religion in the world after Christianity, with 1.8 billion members. However, if the current demographic trends continue, in the second half of this century Islam will surpass Christianity as the largest religious group. The Islamic Declaration on Global Climate Change was adopted by Islamic leaders from 20 countries in Istanbul prior to the Paris Climate Conference and was officially presented at the UN signing of the Paris Agreement on Climate Change in 2015. It called on Muslims around the world and people of all faiths to take urgent climate action to phase out all greenhouse gas emissions by 2050 and commit to 100% renewable energy. The core of the Declaration was a body of ethics known as the Knowledge of Creation (*Ilm ul khalq*) based on the Qur'an, the holy text of Muslims. A central tenet is the definition of humanity's place in Creation as stewards, respecting the "perfect equilibrium" of nature. The Earth is a sacred responsibility entrusted to humans by Allah. Every being exists within an interactive community, glorifying Allah. Destruction or loss of any entity is not only tragic and cruel, but also an offence against Allah.[28]

After affirming the scientific consensus on climate change, the Islamic Declaration called on well-off nations and oil-producing states—those with the greatest responsibility and capacity—to lead the way in tackling climate change and supporting vulnerable communities who are already suffering from climate impacts. Those nations and states must "re-focus concerns from unethical profit from the environment, to that of preserving it and elevating the condition of the world's poor." It ended with the sweeping demand that nations and states "set in motion a fresh model of wellbeing, based on an alternative to the current financial model which depletes resources, degrades the environment, and deepens inequality." It put corporations, finance, and the business sector on the hook to "shoulder the consequences of their profit-making activities, and take a visibly more active role in reducing their carbon footprint and other forms of impact upon the natural environment." Echoing *Laudato si'* and the *Dalai Lama in Conversation with Leading Thinkers on Climate Change*, the Declaration issued the moral challenge: "Realize that to chase after unlimited economic growth in a planet that is finite and already overloaded is not viable. Growth must be pursued wisely and in moderation; placing a priority on increasing the resilience of all, and especially the most vulnerable, to the climate change impacts already underway and expected to continue for many years to come."[29]

Has the Declaration had an impact? According to İbrahim Özdemir, an environmentalist and professor of philosophy at Uskudar University, Turkey, not so much. Although many Muslim majority countries bear the brunt of climate change, their cultural awareness of the crisis and climate action are often "staggeringly limited," he writes. However, he is hopeful that the concept of stewardship that was the basis for the Islamic Declaration on Climate Change can reach the hearts and minds of the almost 2 billion Muslims around the world. He believes that the only effective approach is a "homegrown" environmental movement—Islamic environmentalism—based on Islamic tradition rather than imported "white savior" environmentalism based on first-world political campaigns.[30]

Meanwhile, young Muslim women have stepped forward to honor the Islamic principles. Here are just a few. Kadiatou Balde and Zainab Koli, two young Muslim women from New York, recently founded Faithfully Sustainable, which blends faith and environmental sustainability through education, activism, and entrepreneurship. They reach their online community of more than 2,500 members via social media posts, talks, and webinars, encouraging them to understand and adopt sustainable practices. Ndéye Marie Aïda Ndiéguène, a 24-year-old entrepreneur and civil engineer from Senegal, is the founding CEO of the social enterprise Ecobuilders Made in Senegal. One project uses recycled tires, bottles, and natural materials to build affordable crop storage space for farmers to prevent crop loss and maximize food security; another uses laterite, a red clay-like material and local natural resource, as an eco-friendly building material. Mishka Banuri is a 20-year-old Muslim Pakistani American who co-founded Utah Youth Environmental Solutions to mobilize young people in Utah around the issue of climate justice and hold government officials accountable in combating the climate crisis. She grounds herself in the Islamic principle of the interconnectedness of all living things. In 2018, Banuri helped lead efforts to encourage the Utah legislature to recognize the validity and existential threat of climate change. Utah became the first conservative state legislature to pass a resolution acknowledging the existence of climate change, its causes, and the need for solutions.[31]

Having developed from an agrarian society, Judaism has a direct link to nature and the environment. That link is evident in the Book of Genesis (the first book of the Christian Old Testament and the Hebrew Bible) in which humans are put in the Garden of Eden to be stewards of the Earth. Dozens of Jewish organizations are engaged in climate activities, both in the United States and around the world, especially in Israel. Though as a group

Jews have been active, the various branches hold differing philosophies on the need for activism, with left-leaning Reconstructionist being the most engaged, followed by Reform, Conservative. and Orthodox in that order. This is a relatively small but politically powerful population, numbering fewer than 7 million in the United States and 14.7 million worldwide.[32]

The international Jewish Climate Change Campaign was announced in 2009, at a major interfaith event at Windsor Castle in the UK. It aimed to unite the Jewish people—both the Diaspora Jewish communities and the state of Israel—in addressing global sustainability and the existential threat of climate change. The campaign was a call to action and a call for Jews to be at the forefront of education, action, and advocacy responses to the challenges of climate change and environmental degradation.[33]

Beginning in 1988, numerous Jewish organizations have been founded in the United States to fight climate change including the Jewish Women in Environmental Activism (1988), the Coalition on the Environment and Jewish Life (COEJL, in 1993), and the Jewish Climate Action Network (2013). In 2015, a group of 425 American rabbis, cantors, and other American Jewish leaders and teachers issued a desperate call to Jewish people to act on the climate crisis and to seek eco-justice: "Our children and grandchildren face deep misery and death unless we act. They have turned their hearts toward us. Our hearts, our minds, our arms and legs, are not yet fully turned toward them." They called for a new sense of eco-social justice—a *tikkun olam* ("repair the world") that includes the healing of our planet. They urged those who have been focusing on social justice to address the climate crisis, and those who have been focusing on the climate crisis to address social justice. Another organization, the Jewish Climate Initiative, was founded by Rabbi Jennie Rosenn in 2020 to "stimulate Israel and the Jewish people to play their full role in humanity's efforts to ensure a stable future climate for our children."[34]

However, according to David Dunetz at the Heschel Center for Sustainability in Tel Aviv, which leads the Israel Climate Forum in Israel, in Israel climate change hasn't been at the forefront of politics or the public mind, and the country has lagged far behind in the transition to renewable energy generation, which accounts for just 4% of its energy mix. But there is a growing and broadening climate movement, with different civil society groups stepping up to take on the challenge of climate change—from social rights activists to religious groups and young people who are striking on Fridays. The Heschel Center is working closely with a broad range of stakeholders to develop new climate protection legislation. Dunetz is inspired by Greta Thunberg's Fridays for Future.

Her approach is in many ways the opposite of what I had been preaching all these years as an environmental educator: that we should avoid alarmist and apocalyptic stories and instead try to inspire hope and take the optimistic view that we can do something. And then along comes Greta and explains that our house is on fire and we could very well face extinction. And it's working! And she is right, of course: If we take the science literally that is a real possibility, and governments are not working fast enough. . . . It turns out that fear works as a message—at least to a degree—and the gradualist approach that we practiced on sustainability agendas has not delivered the transformation we need. Young people, it turns out, are ready to hear the truth and act upon it because they understand that this is their future. But as Greta says, this is not the task of young people alone. They need us adults to get into gear and move faster. It has been a vital wake-up call. Hope is essential, but we shouldn't make it a prerequisite for action. Or, as Greta says, "When we start to act hope is everywhere. So instead of looking for hope—look for action. Then the hope will come."[35]

Evangelicals comprise nearly a quarter of the United States population. They are a diverse group drawn from a variety of denominational backgrounds including Baptist, Pentecostal, Reformed, and nondenominational, among others. As a group, Evangelicals generally interpret the Bible literally; they tend to be socially and politically conservative. Most American Evangelicals self-identify as Republicans and view environmentalism and climate change as liberal issues. Among Christians, however, they are the group most likely to believe that God expects humans to be good stewards of nature, and the majority of Evangelicals say it is important to them personally to care for future generations, the natural environment, and the world's poor, and that reducing global warming will help future generations.

The Evangelical Climate Initiative was launched in February 2006, by the National Association of Evangelicals. The statement, Climate Change: An Evangelical Call to Action, has been signed by more than 300 American evangelical Christian leaders including megachurch pastors, theologians, and the presidents of 39 evangelical colleges. In the Preamble, they recognized both their opportunity and their responsibility to offer a biblically based moral witness that can help shape public policy in the most powerful nation on earth and therefore contribute to the well-being of the entire world. "Whether we will enter the public square and offer our witness there is no longer an open question. We are in that square, and we will not withdraw." The statement was frank in acknowledging previous error: "For most of us, until recently this has not been treated as a pressing issue or major priority." Now, however, "Christian moral convictions demand our response to the climate change problem", to follow biblical requirements to care for God's creation, to love others as ourselves, and to exercise stewardship over the earth and its creatures. Notably, the signatories described various specific cost-effective, market-based solutions, available now, that will also create jobs, clean up our environment, make us more efficient, and reduce our dependence on foreign oil, thereby enhancing our national security. The declaration concluded: "The need to act now is urgent."[36]

Economic opportunity has swayed many Evangelicals to support climate action. For example, the majority of Evangelicals are in favor of tax rebates for people who purchase energy-efficient vehicles or solar panels. Katharine Hayhoe, an Evangelical atmospheric scientist and political science professor at Texas Tech University in Lubbock, Texas, says she sees attitudes changing among this group because Evangelicals realize that certain of the solutions to climate change are business-friendly. "There's been a massive expansion of clean energy across the middle of the country. We have 30,000 jobs in Texas with solar energy," she said. "Very conservative farmers, who would rather cut off their arms than let the government impose regulations on carbon emissions, now have wind turbines on their land. And a check arrives in the mail. So they're like, 'oh, ok, there's no problem with that.'"[37]

Climate change has brought together faith communities around the world. The Interfaith Center for Sustainable Development, based in Jerusalem and founded in 2010, works on a global basis to promote interreligious cooperation on environmental sustainability and involve religious leaders, seminary students, and communities in climate change and other environmental issues. One of their recent projects is the faith-based campaign for solar power in Africa, in partnership with Gigawatt Global and Green Anglicans. The project website quotes from the Bible: "And God said, Let there be luminaries in the expanse of the heavens . . . to shed light upon the earth. And it was so". [38]

Interfaith Power and Light is a broad-based coalition of religious interests in the United States founded in 2000 by Reverend Sally Bingham, an Episcopal priest. Its goal, as the name implies, is to respond to global warming by driving smarter energy policymaking and helping thousands of congregations address global warming by modeling energy stewardship in their own facilities. Says Bingham, despite their differences, all major religions share several common principles which compel action on the climate issue. These are the mandate to care for God's creation and the moral obligation to care for the poor and vulnerable. "And if we think of neighbors as the next generation, we have to think about how our behavior is going to affect the people who come after us. And it's kind of sad to think that we may care more about ourselves than we do about our children, and sometimes we behave that way."[39]

THE EXPERTS: SCIENTISTS, DOCTORS, AND ACADEMICS

The messages from scientists have been hugely influential when effectively communicated by the media and amplified by the powerful voices just mentioned. Particularly influential are the widely cited consensus documents of the International Panel on Climate Change (IPCC) that are published about every 6 years. Established in 1988, as part of the UN Environment Programme, the IPCC provides governments around the world with scientific information they can use to develop climate policies. Thousands of highly respected scientists have volunteered their time in writing these comprehensive assessments. These authoritative reports are also a major input into international climate change negotiations, as when the IPCC Fifth Assessment, issued in 2013–2014, provided the key scientific support for the 2015 Paris Climate Agreement that was signed by almost every country.[40]

The IPCC *Special Report on Global Warming of 1.5°C* in 2018 had the greatest impact of any of their reports over the past 30 years. The report was prefaced by the quote from Antoine de St. Exupéry, "Pour ce qui est de l'avenir, il ne s'agit pas de le prévoir, mais de le rendre possible" ("As for the future, it is not a question of predicting it, but of making it possible"). It addressed the impacts of global warming of 1.5°C above the pre-industrial level "in the context of the urgent need to strengthen the global response to the threat of climate change and support efforts to eradicate poverty." The report confirmed once again the direct human influence on the climate system, stating that global warming is likely to reach 1.5°C between 2030 and 2052 if it continues to increase at the current rate. To avoid catastrophic effects of climate change, global warming must not exceed this 1.5°C limit. (Even at that level of warming, the report warned, there would be damaging effects.) The bottom line: avoiding overshoot and reliance on largely unproved technologies for carbon dioxide (CO_2) removal can only be achieved if global CO_2 emissions start to decline well before 2030. From this report came the cry of advocates who believe we have only about 10 years to turn this around. Coming, as it did, from a traditionally cautious and conservative group of scientists, the report had the effect of a deafeningly loud fire alarm, reverberating around the world, particularly among youth, and triggering large climate strikes and protests around the world.[41]

Also influential are the Reports of the Lancet Commissions on climate change and health. The messages of the 2018 Lancet *Countdown on Health and Climate Change: Shaping the Health of Nations for Centuries to Come* were dire, but contained a sprinkling of hope. Trends in climate change impacts, exposures, and vulnerabilities show an unacceptably high level of risk for the current and future health of populations worldwide, the authors wrote. And the lack of progress in reducing emissions and building adaptive capacity threatens both human lives and the viability of the national health systems they depend on. But the report also noted that the health profession was beginning to rise to this challenge and a number of sectors have seen the beginning of a low-carbon transition. Through their frequently cited writings, these and other experts have helped to shape awareness of the urgency of the problem and sounded the imperative to act on behalf of vulnerable groups such as children, certain racial/ethnic groups, and the poor.[42]

I believe that scientists and economists, working together, can play a larger role in motivating action by presenting holistic assessments of the health and related economic costs of the many impacts of fossil fuel combustion on health, particularly for children. These reckonings can then be flipped to show the full benefits of action to reduce or prevent these impacts. As we will read in the next chapter, "Success Stories", researchers including myself and my colleagues have estimated substantial health benefits to children—and associated economic savings—from current climate policies based on the reductions of air pollution. Because valuations of the cost of illness or developmental impairment have generally focused on more immediate, short-term costs, the resulting economic assessments have been serious underestimates. Here is a role for collaborations between health economists and epidemiologists to provide more comprehensive estimates of the potential benefits of averting serious harm to children's health.

Pediatricians are among the most highly respected and trusted members of their communities. They have risen to meet the challenge with a new national network known as the American Academy of Pediatrics' (AAP) Chapter Climate Advocates Program. This grassroots effort aims to raise awareness of the effects of climate change on children's health nationwide. "AAP advocates underscore the idea that if only every parent in America knew that climate action was essential to the well-being of their child and family, we would have no political discourse, no debates about the science, no concerns about the course of action," explained Aaron Bernstein, a pediatrician at Boston Children's Hospital and Harvard University. Now with at least 101 chapters across all states in the United States, the network advocates on behalf of children, works to adopt local climate resolutions, and lobbies for local laws to combat climate change. Several of the chapters have adopted climate resolutions that acknowledge and lay out the impacts of climate change on children's health; others are in the process of passing similar measures.[43]

The network is the brainchild of Lori Byron, a semi-retired pediatrician in Harding, Montana, who worked for three decades as a general pediatrician under the Indian Health Service on the Crow Tribe reservation. She witnessed firsthand the effects of air pollution and climate change: an infant death that she attributed to dirty air and families hit by flooding who lost their homes and were living in a Federal Emergency Management Agency (FEMA) trailer, somebody's spare bedroom, or even their car years after the disaster. Knowing as she did that the home environment is one of the most important social determinants of parents' and children's health, Byron was galvanized into action. In 2018, the AAP chapter in Montana became the first in the country to adopt a climate change resolution, and she began pushing for a broader network of climate advocates in AAP chapters across the United States, each of which responds to the specific environmental challenges facing their local communities. For example, the AAP chapter in Virginia responded to a report that pollution from transportation causes thousands of cases of bronchitis and related symptoms in children by advocating for reform of the state's transportation system. In response to the fact that Virginia's coastal communities are experiencing the highest rate of sea rise along the eastern seaboard, the local AAP chapter lobbied successfully for a law—the first in the South—committing the state to deriving all of its electricity from renewable sources by 2045. In California, Lisa Patel, a pediatric hospitalist and member of AAP, recounted that, in the fall of 2020, she saw in her practice the health effects of the devastating wildfires that hit the region: "It was a quiet summer, but then in September and October, I started seeing a lot of premature labor. . . . These are women that were a month or two early in delivering their babies." In Ohio, concern about allergies, asthma, and preterm births related to extreme heat galvanized AAP advocate and pediatrician Aparna Bole to testify before the Ohio Congress against a bailout of an Ohio coal power company.[44]

Health professionals—physicians, nurses, social workers, psychologists— are being called upon to become more involved in climate advocacy. In a March 2021 article entitled "Health Professionals, the Paris Agreement, and the Fierce Urgency of Now," 13 health professionals and representatives of health organizations wrote that health professionals and health organizations must join the growing global community of science-based advocates working to achieve the goal of the Paris Agreement: "Doing so can be our greatest contribution to the health and wellbeing of all people, especially the world's most vulnerable, marginalized and disempowered people who tend to be harmed first and worst." They used the phrase, the "fierce urgency of now" (most memorably used by Dr. Martin Luther King in his 1963 "I Have a Dream" speech calling for racial justice). They asked health professionals and their organizations worldwide to mobilize with advocacy that resonated across the political spectrum and was adapted to local political, social, and cultural contexts.[45]

In September 2021, ahead of the Glasgow, Scotland, climate talks in November, the editors of more than 230 medical journals said in a joint editorial that the impacts of climate change could become catastrophic and irreversible unless governments do much more to contain global warming to below 1.5°C. The authors also warned that the aim to reach net zero was relying on unproved technology to take CO_2 and other greenhouse gases out of the atmosphere. Simply urging the world and the energy industry to transition from fossil fuels to renewables is not enough: society-wide changes were urgently needed.[46]

In an article in the *New England Journal of Medicine*, Kari Nadeau of Stanford University and I wrote: "This crisis requires that health professionals protect pregnant women and children who are most vulnerable and help them adapt. It also calls upon them to actively engage in efforts to mitigate climate change in ways that are just."[47]

ENVIRONMENTAL ORGANIZATIONS

There are hundreds of environmental organizations around the world—large and small, increasingly working collaboratively to battle climate change. It is impossible to do justice to them all, but I've cited a few to give a sense of their range. The US organization, Moms Clean Air Force, was founded in 2011, with the goal of creating a healthier environment for their children to grow up in. The co-founder, Dominique Browning, a writer, says: "I realized that scientists were only talking to scientists, economists to economists, doctors to one another, and even environmentalists spoke their own language about the issue—so that no one was explaining things to people who wanted to understand but needed explanations that weren't laden with jargon." Moms Clean Air Force describes itself as "the first and largest group to harness mother love to re-engage old-fashioned habits of citizenship." The group has rallied more than 1 million mothers and fathers to join their ranks in the fight for cleaner air and equitable solutions to air pollution and climate change. Through a network of over a dozen state-based community organizers, the moms—avowedly apolitical—meet with lawmakers at every level of government and on both sides of the political aisle to build support for equitable, just, and healthy solutions to pollution. When members marched in the 2017 People's Climate March in Washington, DC, there were people of all ages, from mothers with newborn babies to high school children, from young moms and dads to grandmothers. With wry humor, the Moms team asked marchers to sign "Lunch Box Notes" to their members of Congress to remind Senators and Representatives that "Every day, 77,000 kids miss school because of asthma."[48]

Large environmental organizations—sometimes referred to as "Big Greens"—have played important roles nationally and internationally. They tend to be heavily staffed and well-funded, with annual budgets in the tens of millions of dollars. Among them are the World Wildlife Fund, Natural Resources Defense Council (of which I am a board member), Environmental Defense Fund, Sierra Club, World Resources Institute, European Environmental Bureau, Conservation International, Arab Forum for Environment and Development, Climate Action Network, and 350.org. In recognition of the evolving science and environmental injustice worldwide, many of these organizations have broadened their original focus on conservation to include climate change, public health, and partnerships with environmental justice, Indigenous, and community groups.

350.org was founded in 2007 by American environmentalist Bill McKibben and a group of students from Middlebury College in Vermont. It is probably the best known climate change-focused organization. With its decentralized network extending to 188 countries, it has mounted on-line campaigns, grassroots organizing activities, and mass public actions to oppose new coal, oil, and gas projects; end financing for companies that are heating up the planet; and build 100% clean energy solutions. There is some irony in the name: 350 stands for 350 ppm (parts per million) of CO_2, which has been identified as the safe upper limit to avoid a climate tipping point; however, as of June 2022, the concentration was 421 ppm and going up.[49]

350.org has mounted highly publicized campaigns, such as against the Keystone XL pipeline, large-scale fossil fuel companies, fracking in various cities in Brazil, and grassroots mobilizations before and after the Paris Agreement. 350.org was one of the leading organizers of the Global Climate Strike in September 2019, which took inspiration from the school strikes of the Fridays for Future movement. More than 7.6 million people across 185 countries participated in this mass mobilization event, making the Global Climate Strike the largest climate mobilization in history. At a concert during the Paris Climate Talks in 2015, Bill McKibben took the stage at an event in the Bataclan Theater in Paris that had been the site of a deadly terrorist attack on November 13 that same year. The audience, including myself, had shown up in a major way—there was standing room only—rallying around the climate issue and showing solidarity with the French.

ARTISTS

Poets and other artists have used their power to change minds. In 2017, 19-year-old African American poet and activist Amanda Gorman became the first person to be named the American National Youth Poet Laureate. In her poem "Earthrise" she describes the photo of the earth taken from Apollo 8 as it orbited the moon on Christmas Eve 1968. She then asks us who are "caught in the throes of climactic changes" to "dream a different reality": to make it our mission to "conserve, to protect, to preserve that one and only home that is ours, to use our unique power to give next generations the planet they deserve." Her message is urgent:

There is no rehearsal. The time is
Now
Now
Now.[50]

British artist and filmmaker, John Akomfrah was born in Ghana and lives in London. His work "Purple" is an immersive, six-screen video installation, accompanied by a hypnotic sound score. We see how human encroachment and climate change have drastically altered the physical environment across the planet and their effects on human communities, biodiversity, and the wilderness. The work combines filming in more than 10 countries and archival films to show various disappearing ecological landscapes around the world including in Alaska, Greenland, the Tahitian Peninsula. and the Marquesas Islands in the South Pacific. There are references to acid rain, CO_2, pollution, and war, all showing the fragility of nature and humanity. The color purple in West African countries refers to death. But amid the devastation there are many images of children—on a carousel, in a museum, running and playing—and a video of a baby being born with church bells ringing in the background, all reminding us of our responsibility and signaling hope for the future. In an interview Akomfrah explained, "This is not the 18th century anymore. It is not unlimited landscapes and unlimited space to explore ad infinitum, wasting away, trashing away as we go along. And that sense of game over, of finitude and the encroaching enclosure is the animating impulse behind works like this." The work is about how we have all been complicit in this destruction, but we can do something about it.[51]

YOUTH ACTIVISM AND ELDERS

We now turn to youth activism—the voice of those whose lives and livelihoods are most affected by the lack of urgency on climate change and its co-effects. Youth are making themselves heard, especially in recent years, with coalitions established throughout the world. Examples are the group of youth who brought the *Juliana* case in the United States, Greta Thunberg and Fridays for Future, the International Youth Climate Movement, Zero Hour, and many more.

The Juliana Case

In 2015, 21 children and adolescents between the ages of 8 and 19, including 19-year-old Kelsey Juliana, filed suit against the US federal government charging that the government's inaction on addressing climate change violated their constitutional right to life, liberty, and property. Colleagues from Harvard University and I coauthored an article to support the *Juliana* case by summarizing the scientific evidence on the present harm to children from climate change and co-pollutants. We wrote, "As the Juliana plaintiffs argue—and we agree—climate change is the greatest public health emergency of our time and is particularly harmful to fetuses, infants, children, and adolescents. The adverse effects of continued emissions of carbon dioxide and fossil fuel–related pollutants threaten children's right to a healthy existence in a safe, stable environment." We noted that earlier we had joined with nearly 80 scientists and physicians and 15 health organizations in an amicus brief to help educate the Ninth Circuit about this extraordinary threat. At present, the case is still in the court. It is not without precedent. Courts in the Netherlands and Colombia have recognized the fundamental rights of children to demand that their governments reduce greenhouse gas emissions.[52]

Young Climate Leaders

As Naomi Klein writes, for years governments have been endlessly discussing the problem of climate change and there have been countless appeals from the public referring to the "children." the "grandchildren," the "generations to come" but none of those rhetorical appeals was persuasive enough to politicians and corporations. Now, the children are speaking (and striking) for themselves as part of a global network. Their power, she writes, is that they have not been trained to mask the unfathomable stakes of our moment in the language of bureaucracy and overcomplexity. They understand that they are fighting for the right to live full lives.[53]

The number and diversity of young climate leaders is remarkable. Greta Thunberg is the best known of the youth advocates and an international celebrity. In 2018, at age 15, she launched an international "school strike for the climate" movement called Fridays for Future. This grassroots movement has inspired students all over the world to skip classes mainly on Fridays to participate in peaceful demonstrations demanding that political leaders act to avert climate disaster and transition to renewable energy. From its beginning in August 2018, when Greta stood alone outside the Swedish Riksdag (parliament) holding a sign that read "Skolstrejk för klimatet" ("School Strike for Climate"), the movement has swelled to millions of strikers across 157 countries (Figure 5.2).

On March 1, 2019, students from the global strike coordination group issued an open letter in *The Guardian*: "We, the young, are deeply concerned about our future. . . . We are the voiceless future of humanity. We will no longer accept this injustice. . . . We finally need to treat the climate crisis as a crisis. It is the biggest threat in human history and we will not accept the world's decision-makers' inaction that threatens our entire civilization. . . . Climate change is already happening. People did die, are dying and will die because of it, but we can and will stop this madness. . . . United we will rise until we see climate justice. We demand the world's decision-makers take responsibility and solve this crisis. You have failed us in the past. If you continue failing us in the future, we, the young people, will make change happen by ourselves. The youth of this world has started to move and we will not rest again." Indeed, on March 15, 2019, more than a million students worldwide participated in school strikes across 125 countries, demanding that adults take responsibility and stop climate change. The movement has invited adults to join their ranks.[54]

In 2019, Greta was named *Time Magazine*'s Person of the Year. In 2020, the slender collection of her speeches, *No One Is Too Small to Make a Difference*, was voted by climate activists one of the 10 best books on climate change—right up there with *The Uninhabitable Earth: Life After Warming* by David Wallace-Wells and *This Changes Everything* by Naomi Klein. What explains her actions and her celebrity? In a moving video, Thunberg says that when she was 8 years old she learned about climate

Figure 5.2 Greta Thunberg in front of the Swedish parliament in Stockholm, August 2018. Credit: Per Grunditz/Shutterstock.

change; she could not understand why so little was being done about this disaster, and she became depressed to the point where, at age 11, she stopped talking and eating. She was then diagnosed with Asperger syndrome and selective mutism. After struggling with depression for several more years, she began her school strike in 2018. Thunberg calls Asperger's her "superpower." She says that it allows her to focus intently on a subject and helps her to see things that other people don't see, especially the cognitive dissonance between people saying climate change is important and doing nothing about the problem. She does not care about pleasing people or being liked; she has a brilliant mind and the courage to criticize world leaders for their failure to take sufficient action to address the climate crisis. In a memorable moment, after crossing the Atlantic on a zero-emissions racing sailboat rather than flying to limit her carbon footprint, she attended the 2019 UN Climate Summit in NYC. She stepped off the boat after what must have been a strenuous journey to excoriate the world leaders for their complacency and inaction.

Entire ecosystems are collapsing. We are in the beginning of a mass extinction, and all you can talk about is money, and fairy tales of eternal economic growth. How dare you! For more than 30 years the science has been crystal clear. How dare you continue to look away and come here saying that you're doing enough when the politics and solutions needed are still nowhere in sight. . . . How dare you pretend that this can be solved with just "business as usual" and some technical solutions? . . . You are failing us. But the young people are starting to understand your betrayal. The eyes of all future generations are upon you. And if you choose to fail us, I say: "We will never forgive you."[55]

Greta is far from alone. Carlon Zackhras of the Republic of the Marshall Islands, a Pacific islands group between Hawaii and the Philippines, organizes with #FridaysForFuture and was a part of the youth coalition that spoke at the UN climate conference in Madrid in 2019. There he described the rising sea levels that threaten his country, only 2 meters above sea level, and urged world leaders to take action. He told them that, 2 weeks before his arrival in Madrid, 16-foot waves forced 200 people from their homes. He reminded the audience that, while the Marshall Islands is among the most affected countries, the nation has contributed only 0.00001% of the world's carbon emissions.[56]

Three notable young American activists are Isra Hirsi, Alexandria Villaseñor, and Jamie Margolin. Isra Hirsi, the daughter of Minnesota congresswoman Ilhan Omar, is an environmental justice organizer and a Black Lives Matter (BLM) activist. She co-founded the United States Youth Climate Strike in 2019 and helped organize more than 100,000 young people to strike for climate justice. Alexandria Villaseñor, the 15-year-old climate activist who is co-founder of the US Youth Climate Strike and founder of Earth Uprising, states: "The climate crisis is the largest intergenerational inequality there is." She draws the link between BLM and climate justice: "We know that there is no climate justice without racial justice. The exploitation of Black people is the greatest extractive system of production of all time and in order to heal the planet, we must have Black and Indigenous liberation." Jamie Margolin is a Colombian American who identifies as a Latina Jewish lesbian. In 2017, at age 15, she co-founded Zero Hour, a youth climate coalition that mobilizes young people to take concrete action on climate and environmental justice and "ensure a livable future where we not just survive, but flourish." In 2019, Margolin testified before Congress, alongside other young climate change activists.

The fact that you are staring at a panel of young people testifying before you today pleading for a livable earth should not fill you with pride, it should fill you with shame. . . . People call my generation, Generation Z, as if we are the last generation. But we are not. We are refusing to be the last letter of the alphabet. I am here before the whole country today announcing that we are instead Generation GND: Generation Green New Deal. The only thing that will save us is a new era. It is right here, testifying before you that history is being made. You've heard of the Reagan Era, the New Deal Era, well the youth are bringing about the Era of the Green New Deal.

Her book, *Youth to Power*, is a "how to" guide for student activists: every step from advice on writing, to organizing events, time management, and utilization of media.[57]

Many other groups have joined the global movement. The aptly named Sunrise Movement, formed in 2017, aims to interrupt the sleep of local politicians and get them to "wake up to the climate crisis and do their jobs." Peacefully occupying the office of Nancy Pelosi, Speaker of the US House of Representatives, after the 2018 midterm elections, they berated the Democrats for failing to respond to the climate emergency. They have called on the US Congress to adopt a decarbonization plan similar in ambition to the New Deal.[1] Extinction Rebellion (XR) is an international grass roots movement inspired by the civil rights movement and the suffragettes. Established in 2018 in the UK, with about 100 academics signing a call to action, the movement now has more than 200 branches globally. Using nonviolent civil disobedience to compel government action, XR demands that the UK government declare a climate and ecological emergency, act now to halt biodiversity loss, and reduce greenhouse gas emissions to net zero by 2025, prioritizing the most vulnerable people to create a livable, just planet. Their first act was to occupy key parts of London—the road in front of parliament and the five main bridges over the River Thames—in one of the largest acts of peaceful civil disobedience in the UK in decades. XR Youth is an autonomous wing made up of activists under the age of 30 and centered around climate justice and consideration of the Global South and Indigenous peoples.[58]

The Elders

Older people have joined the young on the front lines. The groups range from the local to the national and the global, but all are about protecting the young and future generations. In the San Francisco Bay area, a group of older women call themselves the Society of Fearless Grandmothers. The founder, Pennie Opal Plant, says their job is to make sure that young people stay safe: "Older women have a special role in these nonviolent direct actions and marches," she says. "We're not afraid to put ourselves between law enforcement and all of the younger people behind us. . . . We have to do everything we can to protect future generations and the entire system of life." Law enforcement personnel have treated them with surprise and respect as when, before the 2018 Global Climate Action Summit, the grandmothers blocked streets so that young protesters could paint a giant mural of climate solutions behind them.[59]

Inspired by Greta Thunberg's School Strike for Climate Fridays, in fall of 2019, then 82-year-old Jane Fonda launched Fire Drill Fridays to protest against the climate emergency every Friday in the streets of the nation's capital. The protests end with civil disobedience, and often, time spent in jail. Her fire engine red coat symbolizes the crisis, harking back to the Dali Lama's "The earth is our home, and our home is on fire." The group demands nothing less than a green new deal, no new production of fossil fuels, the rapid phase-out of existing fossil fuel projects, and a just transition to a renewable energy economy: "It is time for a global reckoning with the fossil fuel industry. We know that oil corporations like Exxon, Shell, Chevron, and BP will throw their weight behind blocking climate action, having captured our democracy and deceived the public for decades about who they are and what they stand for." The movement demanded a reckoning for those who profit at the expense of people and the planet and called on our leaders to act "with the boldness and ambition that the moment calls for." In the first 7 months of the COVID-19 Shutdown and going virtual, the movement reached more than 6 million people.[60]

A global organization, the Elders, has joined forces with the young as well. It was founded by Nelson Mandela (former President of South Africa) on his 89th birthday in July 2007, and it is now chaired by Mary Robinson, formerly the President of Ireland and the UN High Commissioner for Human Rights. The goal Mandela set for the organization was to use their "almost 1,000 years of collective experience" to work on solutions for climate change, poverty, and HIV/AIDS. Among the elders and elders emeriti are Desmond Tutu (Archbishop and Nobel Peace Laureate, South Africa), Jimmy Carter (former President of the United States), Fernando Henrique Cardoso (former President of Brazil), Hina Jilani (pro-democracy campaigner from Pakistan), Gro Harlem Brundtland (former Prime Minister of Norway), Ellen Johnson Sirleaf (former President of Liberia and Nobel Peace Laureate), and Kofi Annan (former UN Secretary General and Nobel Peace Laureate). They have called on world leaders to avert climate disaster, ensure a just transition to a low-carbon economy, and push for innovative new solutions. Their vision: "a world where people live in peace, conscious of their common humanity and their shared responsibilities for each other, for the planet and for future generations."[61]

This diverse group of climate activists teaches us that no one is too small, too young, or too old to make a difference.

CONCLUSION

There are many common themes across these diverse groups—religious, non-religious, Indigenous, non-Indigenous, young, old, black, brown, white, grassroots and grasstops, large and small, professional and non-professional. These are the moral values of stewardship for the earth (our "home"), justice, care for the poor and vulnerable, and responsibility toward future generations. Their messages are realistic, sometimes grim, but without exception hopeful that all is not lost and that solutions can be found. They all speak about the well-being of children, sometimes in personal ways, sometimes as a moral responsibility, sometimes both. Take Jess Housty, the young woman we met earlier from the Heiltsuk Tribal Council in British Columbia: "When my children are born, I want them to be born into a world where hope and transformation are possible." Or the Dalai Lama: "I'm a monk so I have no children, but people who have children have to think about how life will be for them and their grandchildren." Or the Jewish leaders and teachers: "Our children and grandchildren face deep misery and death unless we act." Or the Episcopal priest, Reverend Sally Bingham: "And it's kind of sad to think that we may care more about ourselves than we do about our children, and sometimes we behave that way." Or the pediatrician, Ari Bernstein: "If only every parent in America knew that climate action was essential to the well-being of their child and family, we would have no political discourse, no debates about the science, no concerns about the course of action."[62]

But what of the youth, themselves? As we have seen, they are worried, and they are angry at adults for creating the crisis and doing little or nothing to fix it. They have vowed to take power into their own hands unless the grown-ups act, and adult leaders have often applauded and honored them. In fact, so much so that some young people have complained that the adults are leaving it to the children to fix the problem that they created. The Dalai Lama found the balance: "I really appreciate Greta Thunberg's efforts to raise awareness of the need to take direct action. . . . However, we cannot rest our hopes only on the younger generation. We have to choose political leaders who will act on this issue with urgency." Together, all the voices we have just heard speak are powering transformative action around the world, as we will see in the next chapters.[63]

CHAPTER 6

Success Stories

INTRODUCTION

To those who are overwhelmed by the magnitude of the problem and do not believe that much can be done to avert climate disaster, and to those who would use this argument to justify business as usual, this chapter offers stories of successful policies that have worked to the benefit of health and the economy. This is strong evidence that there are workable solutions to the interconnected problems of air pollution and climate change. Whether framed as an air pollution or a climate mitigation policy, both target emissions from fossil fuel burning and their shared measures of success (benefits) are most often the avoided health effects attributable to air pollution reduction. Most of the studies have focused on adult health benefits and have given scant mention of the benefits to children—this is a gap that my colleagues and I have worked to fill. The chapter presents several case studies that have centered on children but also gives brief examples from the much larger pool of assessments that have focused on adult health. Together, they orient us to the sheer magnitude of health and economic benefits being realized right now from actions to curb fossil fuel emissions.

Two notes before we begin. The numbers presented here are substantial, in some cases staggering, but we must not let them mask what they mean in terms of avoided human suffering. Visualize what it means for a child to be spared the often lifelong burden of asthma: it means not having bouts of struggling to breathe, panicky runs to the hospital or emergency department, and missed days of school because of asthma—burdens that affect parents and children alike. Visualize, too, as a parent, being spared

Children's Health and the Peril of Climate Change. Frederica Perera, Oxford University Press. © Oxford University Press 2022. DOI: 10.1093/oso/9780197588161.003.0006

the sorrow and burden of caring for a child born too early or too small, or a child with a developmental disability.

A number of the studies in this chapter were conducted by me, working with many colleagues at Columbia and other institutions. For reasons of length, it is not possible to list all those who contributed their time, effort, and expertise to the work. The "we" in this section does not do justice to the many dedicated researchers who made the work possible.

BRIEF EXAMPLES

The Paris Climate Agreement

We begin at the highest geographical scale, with an analysis of the global health co-benefits of this historic accord. In 2015, 196 countries adopted the Paris Agreement, a global action plan to tackle climate change. The agreement, as we know, set a framework for limiting global warming to well below 2°C (F), pursuing efforts to limit warming to 1.5°C (F). Considering the avoided costs of premature mortality in adults, an international team of experts calculated that, in 2050, those benefits would substantially outweigh the mitigation cost of achieving these targets. The ratio of health co-benefits to mitigation costs ranged from 1.4 to 2.45, depending on the temperature objectives and equity criteria considered: that is, the health benefits were from 40% to 145% higher than the costs of mitigation.[1]

The European Union Emissions Training Scheme

Moving now to the supranational level, the EU Emissions Trading Scheme is the first and largest international cap-and-trade program for reducing heat-trapping emissions. Launched in 2005, the program operates in all EU countries plus Iceland, Liechtenstein, and Norway. It covers emissions from around 10,000 installations in the power sector, the manufacturing industry, and airlines, amounting to about 40% of the EU's greenhouse gas (GHG) emissions. (We will come to more details on cap- and trade later on.)[2]

The EU Emissions Trading Scheme is credited with an estimated overall reduction of carbon dioxide (CO_2) emissions by 1.2 billion tons between 2008 to 2016 and a 35% decline in those emissions between 2005 and 2019. This reduction exceeded the original target (a 20% reduction in GHG emissions from 1990 levels by 2020). A 2008 report by a coalition of non-profit groups valued the co-benefits to health from cleaner air under the 20% target at as much as €52 billion. They called for a more ambitious target of a 30% reduction in GHG emissions—that would boost the estimated health co-benefits to as much as €76 billion in the year 2020 alone. To date, emissions in the sectors covered (which do not include buildings and transport) have decreased by almost 43% since the 2005 launch. The European Commission has recently proposed legislation to cut GHG emissions by at least 55% before 2030, aiming for climate neutrality by mid-century. A separate emissions trading system would be established for the residential and transport sectors, with the goal of cutting emissions from those sectors by 43% compared to 2005 levels by 2030. To deal with the likely rise in consumer costs, especially affecting vulnerable households and small businesses, the Social Climate Fund would provide €72.2 billion of EU funds to Member States between 2025 and 2032, to be matched by the national governments. While the ultimate shape of new Emissions Trading Scheme is uncertain, the recent disastrous floods in Europe have emphasized the need for radical action and built political impetus.[3]

China

We go now to the national level. In 2013, China issued a national Air Pollution Prevention and Control Action Plan. As the largest developing country in the world, China has experienced extraordinarily rapid economic growth over several decades and, as a result, has suffered from notoriously high levels of air pollution. Heavy reliance on coal combustion, increases in industrial emissions, growth in motor vehicle use, and soaring construction have all contributed. An analysis of national air quality monitoring and mortality data estimated the health benefit of the air pollution policy for the 5-year period 2013 to 2017 in 74 cities in China. Between 2013 and 2017, annual average concentrations of fine particulate matter ($PM_{2.5}$) decreased by 33% and sulfur dioxide by 54% in these cities. As a result of these marked improvements in air quality, there were an estimated 47,000 fewer deaths and 710,000 fewer years of life lost in 2017 than in 2013 in these 74 cities. These impressive numbers reflect both the high pollution levels in China and the size of the population at risk.[4]

In 2011, China began building the foundation for a national program to address climate change by introducing regional Emissions Trading Zone pilots located in Beijing, Shanghai, Tianjin, Chongqing, and Shenzhen, as well as in the provinces of Guangdong, Hubei, and, later, Fujian. More than 252 million people live in these areas. The pilots have been credited with about a 5% reduction of $PM_{2.5}$ in the covered areas, which translated to more than 23,000 avoided adult deaths and 534 fewer newborn deaths per year, with a saving of more than $41 billion annually in gross domestic product.[5]

Building on that success, in February 2021, China began operating a nationwide emissions trading program. The largest in the world, the program covers one-seventh of global fossil fuel CO_2 emissions. The program regulates coal- and gas-fired power plants, accounting for more than 4 billion tons of CO_2—about 40% of national carbon emissions. Its success is therefore considered critical to achieving China's official commitments, which include lowering its CO_2 emissions per unit of gross domestic product by more than 65% from 2005 levels by 2030; peaking its national CO_2 emissions by 2030; and achieving carbon neutrality by 2060. Under the scenario that China peaks its CO_2 emissions by 2025 (which it is on track to do), compared to the reference scenario where there is no peaking of emissions, there would be a 28% reduction in GHG emissions in 2030. As a result, co-pollutants would fall substantially: $PM_{2.5}$ by 15% and nitrogen oxides (NO_x) by 18%. The estimated health co-benefits under this scenario add up to 23.6 million avoided cases of adult mortality, hospitalizations, and emergency department visits, with a total health benefit valued at $90 billion.[6]

United States

The federal Clean Air Act, passed in 1970 and revised in 1990 with the Clean Air Act amendments, is considered a milestone in public health policy to curb air pollution. Aggregate national emissions of $PM_{2.5}$, NO_x, and four other common pollutants regulated under the Clean Air Act were reduced by an average of 73% from 1970 to 2017. The dramatic improvement in air quality has brought vast benefits. For 2020 alone, the estimated numbers of avoided deaths and illnesses included 230,000 premature adult deaths, 280 infant deaths, 200,000 heart attacks, 66,000 hospital admissions for respiratory conditions, and 2.4 million asthma attacks, as well as many millions of lost days of school and work. These health benefits were valued at almost $2.0 trillion in 2020. Illustrating the false dichotomy of environmental health and the economy, between 1990 and 2020, the US gross domestic product grew by more than 250%. The US Environmental Protection Agency (EPA) determined that the monetized health benefits of the Clean Air Act far exceeded the implementation costs by a factor of 32.[7]

Stockholm

Stockholm is the capital of Sweden and the most populous urban area in Scandinavia. One million people live in the Stockholm municipality, 350,000 reside in the central city, and more than 200,000 workers enter the city every day for work. For 7 months in 2006, the city of Stockholm tried out a congestion pricing zone to reduce traffic entering the central city. According to government estimates, inner-city traffic was reduced by around 20–25% during the trial. Encouraged by the success of the trial program in lowering levels of nitrogen dioxide (NO_2) and PM, the Swedish government decided in 2007 to make the program permanent. As a result of the permanent congestion pricing zone program, NO_2 levels fell by 15–20% and PM levels by 10–15%. Urgent visits and hospitalizations for asthma among children aged 0 through 5 were reduced by 50%. GHG emissions were also lowered.[8]

In-Depth Case Studies

We now turn to case studies that focus mainly on the benefits to children, so often neglected.

The United Kingdom: London

London has suffered from poor air quality since the thirteenth century, with the most horrific air pollution event in UK history occurring in December 1952. The "great smog of London" that took the lives of 12,000 Londoners from respiratory illness was due to coal burning in domestic units, power plants, and factories. The transition away from coal as the city's primary heating source took many years. Although these severe fogs have not occurred since and air quality in London has continued to improve, air pollution remains a challenge (Figure 6.1).[9]

In recognition of the peculiar challenges the city faced, the London Mayor and the Greater London Authority, that together comprise "City Hall", were given control over air pollution and transportation. In 2003, they began taking steps to reduce fossil fuel pollution—particularly from transportation, which had been identified as the major pollution source. Initially, benefits were tallied in terms of decline in air pollutant levels and deaths avoided; children and climate change benefits were a sidebar. Over the years, there has been an increasing focus on protecting children's

Figure 6.1 London: Big Ben and House of Parliament in fog.
Credit: Richard J. Ashcroft/Shutterstock.

health, and climate change mitigation has achieved greater prominence. The pollutants of primary interest have been PM and the more traffic-specific pollutant, NO_2.

Under the first two mayors of London, Ken Livingstone (2000–2008) and Boris Johnson (2008–2016), the pace was slow and the approach piecemeal. Under Mayor Sadiq Kahn (2016–), recently elected to another term, City Hall mounted a more forceful, holistic, and effective campaign to reduce air pollution and GHG emissions. Their goal: to make London a zero emission city by 2050.

To recap progress prior to Sadiq Khan's election, in 2003, Livingstone introduced the congestion charging zone (CCZ) which placed a daily fee on drivers entering central London. This was followed by the Low Emission Zone (LEZ), which applied a charge to all heavy-duty vehicles that did not meet certain emissions standards. The annual net economic benefit in London of the CCZ was estimated to be £122 million (at 2005 price levels), with a ratio of costs to benefits of 2.27. However, in 2008, as Mayor Livingstone's tenure ended, more than 4,000 premature deaths among Londoners were being attributed to long-term exposure to airborne $PM_{2.5}$. Residents in inner-city areas were breathing the most polluted air.[10]

The next mayor, Boris Johnson (2008–2016), took further measures to reduce air pollution and carbon emissions, promoting new hybrid and zero-emission buses, adapting buses to make them cleaner, encouraging cycling and walking, and promoting zero-emitting electric vehicles. However, the estimated toll in 2010 from exposure to $PM_{2.5}$ and NO_2 was over 9,400 premature deaths along with more than 2,000 hospital admissions for respiratory illness; the cost was as high as £3.7 billion. But there was also some positive news with the prediction that, between 2008 and 2012, the decreases in $PM_{2.5}$ and NO_2 levels would result in a gain of more than 580,000 extra years of life in the London population.[11]

Signaling an increasing concern for children's health, experts at King's College London estimated that, between 2014 and 2016, about 1,000 hospital admissions for child asthma—10% of all child asthma admissions in London during this 2-year period—were attributable to air pollution. Looking at the other side of the coin, the team later calculated that if London's air pollution were cut to 20% of then current levels, 7,900 fewer children would experience low lung function and 138 cases of low birth weight would be avoided every year.[12]

In 2013, a municipal audit revealed that low-income communities and communities of color were being exposed to about 20–25% more NO_2 pollution than the city average. According to an air quality report that was completed in 2013 but was not published by Boris Johnson while he was mayor, 24% of primary schools in the capital were located in areas that exceeded EU limits for NO_2 pollution—and four-fifths of those were in deprived areas.[13]

As a first-time candidate back in 2016, Sadiq Khan ran on a platform that prioritized improvements to air quality; he vowed to represent all Londoners and to tackle inequality in the capital. He himself did not come from a privileged start, being one of eight children born to Pakistani immigrants—a bus driver and a seamstress—on a south London housing estate. Having taken stock of all these reports on the toll of fossil fuel pollution on health—and particularly the risks to children and the disadvantaged—he began a series of aggressive actions to curb air pollution and climate change, with attention to environmental and social injustice.

He started by adding a "toxicity charge" (T-Charge) of £10 applied to older, more polluting vehicles in central London, in addition to the £11.50 Congestion Charge fee, to take effect in 2017. The toxicity charge reportedly reduced NO_2 pollution by 36% in the central zone in its first 2 years. In 2019, Sadiq Khan went further, launching the more stringent "ultra low emission zone" (ULEZ) that superseded the toxicity charge. The ULEZ applied to all motor vehicles in Central London. Vehicles that did not conform to emissions standards for their vehicle type would be charged; others could enter the controlled zone free of charge. Recognizing that switching to less-polluting vehicles imposes a heavier burden for low-income and disabled residents and small business owners, the mayor created a fund of £60 million to help these individuals scrap their older cars and replace them with models that met ULEZ requirements.[14]

The ULEZ was expected in its first year to reduce NO_x emissions by 45% in central London, 40% in inner London, and 30% in outer London. After only 6 months NO_x emissions from road transport in the central zone were reduced by 29% compared to a scenario where there was no ULEZ. ULEZ policies were predicted to result in the avoidance of 300,000 new cases of NO_2- and $PM_{2.5}$-related illness as well as more than 1 million new air pollution-related hospital admissions London-wide by 2050. This equated to a 25% reduction in total new cases of disease—including in children—related to air pollution in London.[15]

In addition to the transportation charge programs, the Khan administration aimed directly at pollution from bus and taxi fleets, committing more than £300 million to retrofitting thousands of London buses and phasing out pure diesel double deck buses as of 2018. New taxis licensed in London after January 1, 2018, were required to be zero emission capable. They would be supported by a network of rapid electric charge points, many dedicated exclusively to London's black cab fleet. By 2020, the entire bus fleet had been retrofitted or replaced to meet ULEZ standards, the bus fleet had been substantially electrified, London had 5,000 charge points with over 300 rapid charge points, and the amount of protected space for cycling had almost tripled.[16]

To protect children most affected by transport pollution, early in its tenure the Khan administration initiated air quality audits at 50 schools and 20 nurseries in the most polluted areas of London and subsequently expanded the audit program. The administration identified 12 "hotspot" areas where Londoners, including thousands of schoolchildren, were exposed to some of the highest levels of NO_2 pollution. These were designated Low Emission Bus Zones, into which the greenest buses were to be placed, a move that lowered NO_x emissions by 90% across the routes operating in hotspot areas. Taking another tack as well, the administration created the School Streets program, in which roads surrounding schools are closed to motor traffic at the start and end of the school day (Figure 6.2). The goal was to improve road safety and encourage more families to switch to walking, cycling, or scooting, thereby helping to improve air quality and reduce congestion. Cleaner air and more physical activity would benefit children's health and academic attainment. There are now over 300 School Streets, with hundreds more on the way.[16]

Figure 6.2 The Mayor of London, Sadiq Khan, meets children at Corams Fields Youth Centre.
Credit: Alamy/AP.

Despite all these actions, in May 2018, and 2 years into his tenure as Mayor, Sadiq Khan stated a bald truth.

> Toxic air, noise pollution, the threat to our green spaces and the adverse effects of climate change all pose major risks to the health and wellbeing of Londoners. . . . Big problems simply demand ambitious responses. Nowhere is this truer than with air pollution in London. Thousands of Londoners die prematurely every year as a direct consequence of our dirty air, which often breaches legal limits. Air pollution has been linked to asthma, strokes, heart disease and dementia— and is to blame for children in parts of our city growing up with stunted lungs. Worryingly, some of the worst pollution hotspots are around schools. And research shows that London's most deprived communities are among the hardest hit—meaning that poverty and pollution are combining to limit the life chances of too many young Londoners.[17]

In 2020, the Mayor issued a more upbeat report assessing the effectiveness of his policies in improving the air for Londoners, especially children. The report based its assessment of the improvements in London's air quality between 2016 and 2020 using data from London's air quality monitoring network and modeling from the team at King's College London (now at Imperial College London). It linked the improvements to the many actions taken by the Mayor of London and Transport for London. The overall finding was that, in his first 4 years as mayor, actions by the city government resulted in dramatic improvements in London's air quality, especially for NO_2, and a sharp reduction in the numbers of schoolchildren and Londoners who were living in areas with dangerously polluted air. Among the specific improvements were a 94% reduction between 2016 and 2020 in the number of Londoners living in areas with illegal levels of NO_2 and a 97% drop in the number of schools in highly polluted areas—from 455 in 2016 to only 14 in 2019. Whereas more than 2 million people in the capital had lived with illegal levels of NO_2 in 2016, this number dropped to 119,000 in 2019. Due to the ULEZ, in February 2020, concentrations of NO_2 at roadside sites in the central zone were 44% lower than 3 years earlier, with more than 44,000 fewer polluting cars being driven in the central zone daily. It was estimated that at the rate of improvement in NO_2 levels during the years 2008 to 2016 when Boris Johnson was mayor of London, it would have taken 193 years to reach legal compliance for NO_2. Dr. Gary Fuller, an air pollution expert at Imperial College London, said of the report: "The changes in NO_2 in central London and along main bus routes were some of the fastest that we've ever measured in 30 years of monitoring. . . . These successes show that our city's air pollution is not an intractable problem." By the end of 2019, the ULEZ had reduced CO_2 emissions by 12,300 tonnes, a reduction of 6%.[18]

Taking stock, the Mayor said: "I'm pleased that Londoners are breathing cleaner air and that we're saving the National Health Service billions of pounds. . . However, air pollution remains a major public health challenge and it's time for the government to step up."[19]

In October 2021, the ULEZ was expanded to an area 18 times larger, covering all of inner London. The ULEZ expansion, alongside tighter London-wide Low Emission Zone standards for heavy vehicles introduced in March, is expected to have reduced NO_x emissions from road transport by 30% in 2021, covering a population of 4 million. This would be accompanied by a 5% reduction in CO_2 emissions, helping to push the city toward net zero emissions by 2030. Although the total health benefits of London's emission zone policies have not been estimated, many thousands of children have been spared being born preterm, suffering asthma attacks and asthma hospitalizations, and developing asthma, other respiratory diseases, and other developmental problems.[20]

Air pollution is not the only challenge facing Londoners. The Mayor's 2018 London Environment Strategy had noted the evident truths that the city remains over-reliant on the fossil fuels that are a major contributor to climate change and that London is not yet on track to reduce its emissions quickly enough to avoid the worst impacts of climate change or to meet national and international climate goals. London's GHG emissions are dominated by office and residential buildings (76%) and transport (24%). Rather than taking a piecemeal approach, the mayor set out a holistic plan for London to be a zero-carbon city by 2050, with energy-efficient buildings, clean transport, and clean energy. This is a stunning challenge, especially since, by 2050, London is expected to be home to more than 11 million people, compared to around 8.7 million people today. The growth in the size of the population and economy will pose ever greater environmental challenges, but the Administration has expressed confidence that new thinking and careful planning can provide the solution to many of the threats the city now faces. City Hall has developed an implementation plan detailing the specific steps that need to be taken by 2023 to translate each of the mayor's policies to action, with attention to environmental justice and the need to support the most vulnerable and disadvantaged Londoners.[21]

In conclusion, Mayor Sadiq Khan and his colleagues in City Hall deserve praise for their achievements in protecting the health of Londoners, especially children and the less advantaged. I believe that the Mayor's focus on children was a key factor in the political success of the policy. Also critical was his commitment to tackle inequality in the capital while emphasizing the benefits to all Londoners. The candor and directness with which Mayor Khan told the public that, despite successes, the vast scale of the harm from air pollution and climate change demanded even more ambitious responses were certainly unusual in a politician seeking re-election for a second term and probably swayed many voters to support him.

London has been recognized as a leading example of how cities can transition to sustainability, becoming one of five finalists for the prestigious 2020 World Resources Institute's Ross Center Prize for Cities. The public has also recognized the achievements: in May 2021, they re-elected Sadiq Khan for a second term.[22]

Krakow, Poland

In December 1991, I stepped off the train that had brought me from Warsaw to Krakow in the Southern coal-rich region of Poland. In my search for a heavy air pollution area for a study of the effects on the developing fetus, I had been introduced to Wieslaw Jedrychowski, a professor of epidemiology and pulmonary physician at the Jagellonian University in Krakow, the second oldest university in Central Europe. Wieslaw had invited me to come for a visit to plan a study. As I stood on the platform, grayish flakes were spiraling downward. Puzzled, I stared upward, wondering what they could be . . . until I felt some water droplets on my still warm face. These were, of course, melting snowflakes—albeit snowflakes darkened by soot. I knew right away that this would be the perfect place for a study in pregnant women and their children of the health impacts of the air pollutants I had been researching for a number of years: the polycyclic aromatic hydrocarbons (PAHs) found in airborne PM.

At the time, Krakow had the dubious distinction of being one of the most polluted cities in the world due to the intensity of coal burning, with an accompanying high rate of respiratory and other health problems in children. Figure 6.3 shows Krakow under a blanket of smog. We needed a fuller picture of the consequences of air pollution exposure on children's health. Although much higher than in the United States, air pollutant levels in Krakow were within the range seen in other coal-burning cities in Europe, China, and India. We anticipated that our findings would therefore be useful for campaigns in other areas of Europe and in China and India,

Figure 6.3 Smog over the city of Kraków, Poland.
Credit: BeeZee Photo/Shutterstock.

which also were struggling with poor air quality and often unresponsive governments.[23]

Wieslaw Jedrychowski and I met over the next week and scoped out the first phase of the research. Back home, a then doctoral student at Columbia, Robin Whyatt, and I worked with the Polish team to put together a proposal to a foundation; it was funded, and we were off. Over the next year, we enrolled 160 newborns from Krakow and Limanowa, both in Southern Poland and with varying levels of exposure. We measured PAH-DNA adducts (a biomarker of PAH exposure during pregnancy) in maternal blood and newborn cord blood collected after delivery. Newborns with high levels of PAH-DNA adducts, signaling higher exposure during their months in the womb, had significantly decreased birth length, weight, and head circumference. This initial study provided the first molecular evidence that transplacental PAH exposure to the fetus compromises fetal development. We determined to mount a much larger long-term study to learn about the consequences of prenatal exposure to PAHs and $PM_{2.5}$ on children's respiratory health and cognitive function.[24]

This gleam in the eye led, in 2000, to the launch of a long-term prospective cohort study of more than 500 children of nonsmoking mothers in Krakow who were enrolled during pregnancy. We monitored mothers' personal exposure during pregnancy to PAHs and $PM_{2.5}$, as well as exposure inside and outside the home, and we repeated the home monitoring when the children were 3 years of age. Over the next years, birth outcomes, children's respiratory health, and cognitive and behavioral development were carefully tracked. As we had predicted, high prenatal PAH exposure was associated with reduced birth weight and head circumference, lower child intelligence, and a range of neurobehavioral problems in children. In addition, prenatal exposure to both PAHs and $PM_{2.5}$ were each associated with increased risk of respiratory problems, including coughing and wheezing in the early years and airway inflammation. Prenatal $PM_{2.5}$ exposure was also associated with increased risk of difficulty breathing, decreased lung function, and recurrent bronchopulmonary infections. We found that maternal psychological distress and high PAH exposure experienced concurrently by the mother during pregnancy combined to increase the risk of anxiety, depression, and other neurobehavioral problems in the children seen at age 9. We have continued to follow the children as they moved through adolescence to learn about further impacts of air pollution on their physical and mental health.[25]

From the early years to the present, this research has been instrumental in prompting the Krakow government to take action to curb coal burning, the dominant source of air pollution. The pace of government response was slow at first. In 1995, the Polish government and the government of Krakow began to offer a co-financing program that lasted through 2009 to help residents replace coal stoves and boilers with natural gas or electricity. Although there were tens of thousands of conversions throughout the program's duration, several years later 65,000 residential coal stoves and small coal-fired, low-efficiency boilers and 2,800 residential boilers using solid fuels were still operating in the city of Krakow. A study in 2009 confirmed that residential heating by coal combustion in small stoves and boilers continued to be the major contributor to the extreme pollution levels in Krakow—especially during the heating season—with road transport and industry playing a lesser role. It was clear that the government of Krakow had failed to adequately address the problem.[26]

After that, efforts to reduce air pollution intensified. Citing the health risks, citizen movements such as the Smog Alert Campaign (founded in 2012) upped the pressure on national and local legislators to take action. They also organized educational campaigns to focus people's attention on the health impacts of polluted air that increased awareness of the problem among citizens. On several occasions, Wieslaw and I spoke in radio interviews and briefed Smog Alert leaders about the health impacts we were seeing. In March 2012, the European Commission referred Poland to the European Court of Justice, asking the Court to impose large fines on Poland for air quality violations. By March 2015, a total of 54 coal-reducing programs had been implemented in Poland overall. Finally, in January 2016, the Parliament of the Małopolska Province, of which Krakow is the capital, adopted a resolution banning the use of solid fuels.

Having reported the "bad news" for so many years, we now had a welcome opportunity to track the effectiveness of these government initiatives, both in our cohort and in the city as a whole. Following Wieslaw's death in 2015, the team at the Jagellonian University in Krakow has been ably headed by Professor Agnieszka Pac. The first step was to record trends in levels of air pollution over time and create a timeline of the major policies and regulations implemented between 2010 and 2020 to reduce coal-burning pollution in Krakow. The next step was to determine the extent to which the trends in levels of $PM_{2.5}$ and PAHs corresponded to that timeline (they did). Then we were ready to estimate the health benefits in children due to the change in pollutant levels.

The results showed clear benefits. First, in our cohort study, repeat air monitoring during the heating season (October to April) at the same children's homes showed a significant reduction in $PM_{2.5}$ between 2001 and 2016. There was also a significant downward trend in city-wide levels of $PM_{2.5}$ measured during successive heating seasons between 2010 and 2019.

As we see from the graph in Figure 6.4, the concentration of $PM_{2.5}$ in Krakow during the heating season dropped by 50%: from roughly 60 μg/m³ in 2010 to 30 μg/m³ in 2020, indicating that the policies to control coal burning for home heating were effective in bringing down the concentrations of fine PM in Krakow. The average annual concentrations were also markedly lower in 2019 compared to 2010: dropping from 40 μg/m³ to about 25 μg/m³. Nevertheless, the $PM_{2.5}$ levels in 2019 were still more five times higher than now recommended by the World Health Organization, so more needs to be done.[27,28]

To quantify the city-wide benefits for the health of children and adults from reductions in $PM_{2.5}$, we used the US EPA's Benefits Mapping and Analysis Program (BenMAP). This publicly available computer program was developed to estimate the number of air pollution–related illnesses and deaths and their associated economic values. We adapted the BenMAP software to the Krakow setting by including geographic, demographic, and health data for Krakow, as well as concentration-response and cost functions most relevant to the Polish population. In addition to the traditional adult health outcomes (i.e., premature adult deaths, adult hospitalizations, and emergency department visits), we estimated the

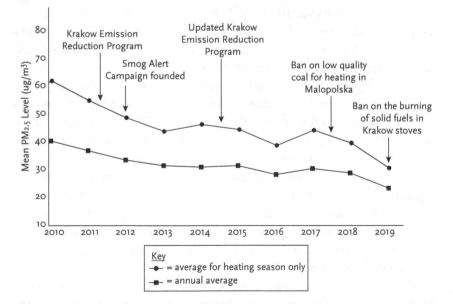

Figure 6.4 Average fine particulate matter ($PM_{2.5}$) during the heating seasons (2009–2010 to 2019–2020) and the annual average $PM_{2.5}$ concentrations in relation to clean air policies (2010–2020)[27].

Credit: Dr. Agnieszka Pac and Renata Majewska of the Jagiellonian University Medical College in Kraków, Poland.

avoided cases of preterm birth, low birth weight, child respiratory illnesses, and child asthma hospitalizations associated with the change in annual PM$_{2.5}$ concentrations.

Estimated benefits included 320 fewer premature adult deaths (a 5% reduction) and 241 avoided adult emergency hospitalizations from cardiovascular problems attributable to the reduction in the annual average level of PM$_{2.5}$ between 2016 and 2019 (data were available only for those years). Taking the drop in PM$_{2.5}$ between 2010 and 2019, there were substantial gains for children including 77 fewer cases of preterm birth (a 16% reduction), 52 fewer low-birth-weight cases (a 12% reduction), 22 avoided cases of hospitalizations due to upper respiratory infections among children 1–4 years old, and 143 fewer hospitalizations for asthma among children 0–18 years old (a 36% reduction). The associated economic savings were substantial.

From the beginning so many years ago, we intended the research to be a catalyst for action on air pollution and climate change, both in Poland and in other countries relying heavily on coal. Over the years, data from this research have informed parents, public interest groups such as Smog Alert, and local and national government officials in Poland of the harm that coal burning was inflicting on their children from the time they were in utero all the way into adolescence. We were able to show the direct benefits for air quality and health coming from the increasingly tough policies that were implemented to address the problem.

Today Krakow is recognized as a pioneer in tackling both air pollution and climate change, having instituted Poland's first ban on burning coal in homes. The city is using drones to check chimneys for signs that household furnaces are not in compliance. As of January 2020, none has been detected. Although air pollution in Krakow is still too high, the city has been described as an "oasis" of cleaner air relative to the surrounding districts that still belch out pollution from coal burning.[29]

The Clean Air Program has given Krakow an appetite for tackling climate change. Knowledge of the immediate local health benefits of cleaner air coupled with new data showing that the city emits an estimated 5.5 million tons of carbon dioxide each year from transport, buildings, heating, and electricity has prompted Krakow city leaders to join 14 other European cities taking part in a program supported by the European Institute of Innovation and Technology to set ambitious goals to become both carbon neutral and inclusive. As everywhere, youth in Krakow are loudly demanding action to counter climate change. In March 2019, hundreds of young people in Krakow joined the worldwide School Strike for Climate, which included more than a million participants in over 120 countries.

Marching on Krakow's Main Square, the youth held up such signs as "Climate Is Changing: Why Aren't We?"[30]

United States: The Regional Greenhouse Gas Initiative Test Case

In 2018, searching for a regional climate initiative that would lend itself to a fuller assessment of the health benefits to children than hitherto attempted, my eye landed on the Regional Greenhouse Gas Initiative (RGGI), colloquially referred to as "Reggie."

Before we delve further into RGGI, we need a proper introduction to the cap-and-trade systems briefly alluded to earlier. This approach to decarbonizing the economy and battling climate change has won out over a carbon tax in the United States because there is less resistance on the part of citizens and industry to a market-based strategy than to an overt tax. According to the World Resources Institute, each has advantages and disadvantages, but, if well-designed, either a carbon tax or a cap-and-trade program can be an effective centerpiece of climate policy.[31]

How does cap and trade work? In a cap-and-trade system to mitigate climate change, the government sets an emissions limit ("cap") and issues a quantity of emission allowances consistent with that cap. The cap is calculated to put the country, region, or state on a path to meeting its target to cut emissions by a certain amount and by a certain date. Businesses can either lower their emissions or obtain permits ("allowances") from the government to pollute. The government distributes the allowances to the companies (each for 1 ton of GHG emissions) generally through an auction. Companies may buy and sell ("trade") allowances, thus establishing the price of carbon within a market. Companies able to reduce their emissions faster and at a lower cost may sell any excess allowances to companies facing higher costs. The theory is that, by reducing the cap over time, the government provides an incentive for industry and businesses to reduce their emissions efficiently.

RGGI was the first regional cap-and-trade program in the United States designed to reduce GHG emissions from the electric power sector. Established in the northeastern region in 2005, through a Memorandum of Understanding between the governors of Connecticut, Delaware, Maine, New Hampshire, New Jersey, New York, and Vermont, the program was expanded in 2007 to include Maryland, Massachusetts, and Rhode Island. (New Jersey left the program in 2012 but rejoined in 2020.) RGGI was implemented in 2009. Participating states were expected to reduce their annual CO_2 emissions from the power sector by 45% below 2005 levels by 2020, and by an additional 30% by 2030.[32]

RGGI required fossil fuel power plants with capacity greater than 25 megawatts to obtain an allowance for each ton of CO_2 they emit annually. (A typical coal plant is about 600 megawatts in size.) Power plants within the region may comply by purchasing allowances from quarterly auctions or from other generators within the region. They may also offset their emissions by projects elsewhere. RGGI held its first auction for emissions allowances in 2008. The revenues have been reinvested to accelerate renewable energy use, improve energy efficiency, and support other public benefit programs.[33]

Although RGGI was designed to reduce GHG emissions, it also reduced co-emissions of toxic pollutants, including $PM_{2.5}$, NO_x, and sulfur dioxide. Not only is $PM_{2.5}$ emitted directly by fossil fuel power plants, but the gases NO_x, sulfur dioxide, and volatile organic compounds react in the atmosphere to form "secondary" $PM_{2.5}$. Directly emitted $PM_{2.5}$ accounts for approximately 10–70% of all $PM_{2.5}$ in the United States depending on the source (Figure 6.5).[34]

RGGI presented a good first test case for an assessment of the potential benefits of a climate initiative that focused on children. Over the preceding year, my colleagues and I had prepared the groundwork for such an opportunity, scouring the literature for data on the damage functions

Figure 6.5 The last big coal-burning power plant in Somerset, Massachusetts, one of the region's heaviest polluters, shut down permanently in 2017.
Credit: AP Images/Shutterstock.

or "concentration-response functions" that related a specific level of $PM_{2.5}$ exposure to preterm birth, low birth weight, respiratory illness, and autism spectrum disorder in children. These outcomes are known to be or "likely to be" causally related to $PM_{2.5}$ exposure. We published the results of this exhaustive review in a leading journal in the hopes that others would add it to their toolboxes for benefits assessments. Next came a similar extensive review to identify the economic cost of each case of illness or death. This, too, was published in a scientific journal. We had at hand the calculated annual changes in ambient $PM_{2.5}$ concentrations for each county in the region from the previous analysis. These three publications made a strong foundation for our new analysis.

With these key inputs we expanded the publicly available US EPA BenMAP to estimate the child health benefits of $PM_{2.5}$ reduction in the RGGI states and in the neighboring states of New Jersey, Pennsylvania, Virginia, and West Virginia. BenMAP already had the population data for each county, and it only remained for us to upload the new concentration-response functions and background incidence data for each of our new outcomes.

An analysis in 2017 had estimated that, between 2009 and 2014, RGGI had prevented up to 830 early deaths among adults, 9,900 asthma attacks in adults and children combined, 16,000 respiratory illnesses in adults and children, 390 non-fatal heart attacks, and up to 47,000 lost workdays. Avoidance of these and other adverse impacts resulted in an estimated $5.7 billion in health savings and other benefits for northeastern states. Our subsequent analysis added further health benefits to children, estimating that there were 112 fewer preterm births and 56 low-birth-weight babies, 537 avoided new asthma cases, and 98 avoided cases of autism spectrum disorder between 2009 and 2014 as a result of the policy. The economic value of these additional prevented cases was between $191 and $350 million, depending on assumptions of severity and persistence of the condition beyond childhood. There was considerable variation across RGGI and neighboring counties in the distribution of economic benefits from the avoided health impacts in children, with the more densely populated areas receiving higher benefits. This was expected because of the larger at-risk population (more pregnant women and children) breathing cleaner air in these areas.[35]

The estimated avoided cases and associated savings are undercounts. First, we only accounted for the impacts of secondary $PM_{2.5}$, omitting primary $PM_{2.5}$ that is directly emitted. Nor did we consider the health effects of other toxic pollutants like ozone or NO_2. And, as noted earlier, most of the published cost estimates we relied on did not consider the economic loss from long-term effects of early illness or impairment. An example is the impact of a child's cognitive disability resulting from being born pre-term on parents' employment as well as on the child's earning power as an adult.

Despite its limitations, our assessment demonstrated that RGGI has provided considerable child health benefits to participating and neighboring states beyond those conventionally considered. Moreover, those health benefits have significant economic value. Although they are undervalued and dwarfed in economic terms by the traditionally analyzed adult outcomes, the results put a child's face on the benefits of curbing fossil fuel pollution, a factor hitherto missing.

Looking back, RGGI has been a success by many measures. The program has driven significant reductions in CO_2 emissions from the electric power sector. By 2017, it exceeded its initial goal of a 45% reduction in annual CO_2 emissions (below 2005 levels) by 2020. Over the first 10 years of the program, CO_2 emissions from the electric power sector in RGGI states dropped by 47%, from 133 million tons to 70 million tons; these reductions exceeded those of the rest of the country by 90%. RGGI states are well on track to meet their goal of an additional 30% drop by 2030. While RGGI has not been the single driver behind the northeastern region's decarbonization in the electric power sector, the program has been a significant factor in the reduction of CO_2 emissions.[36]

Once again illustrating that a clean environment is good for the economy, a review in 2019 found that the participating states maintained steady growth in regional gross domestic product, outpacing growth in the rest of the country by over 30%. In addition, RGGI states have generated $3.2 billion in allowance auction proceeds, the majority of which have been invested in energy efficiency, clean and renewable energy, and GHG abatement. The cumulative benefits of all investments made through RGGI up to 2018 totaled more than $12 billion in lifetime energy bill savings. The health and productivity benefits of RGGI from reductions in $PM_{2.5}$ levels have been pegged at a hefty $5.7 billion.[37]

Other metrics point to success: as of 2020, 198 of the 203 power plants (almost 98%) subject to RGGI requirements have met their compliance obligations. In 2005, there were 41 coal plants across the RGGI states. By 2020, this number was reduced to 24 (a 41% drop). For non-RGGI states, there was only a 29% reduction in the number of coal plants. Between 2009 and 2020, coal consumption for electricity generation in regulated states dropped by 92%, compared to 52% across unregulated states.[38]

Along with the transition away from coal, RGGI has helped northeastern states achieve major reductions in pollution from electricity, and the states are now headed to a 65% reduction in electricity emissions by 2030. RGGI has also generated funds for investments in efficiency and clean energy that themselves have reduced emissions while saving consumers money and creating new business opportunities in participating states.[39]

RGGI has not been an unalloyed success, however. Despite stated commitments and actions by states in the RGGI region to direct climate change benefits to disadvantaged communities, 29% of facilities regulated under RGGI actually increased their emissions between 2000 and 2019, creating hotspots of higher CO_2 and co-pollutant emissions in nearby neighborhoods. Hotspot neighborhoods had a higher proportion of residents of color and households below the poverty line than those where emissions did not increase. These findings argue strongly for reform of RGGI to guarantee that a much larger share of the health and environmental benefits, as well as the RGGI proceeds, reach these communities.[40]

A compelling argument has also been made that a more ambitious RGGI cap for 2030 is not only feasible given the progress to date, but necessary if the region is to achieve its economy-wide GHG reduction goals. A 10-year review in 2019 estimated that, for the northeastern states to achieve a 45% reduction in economy-wide GHG emissions by 2030 (from 2015 levels), the region would need to achieve a 57% reduction in GHG emissions from the electric sector by 2030. This would yield a 2030 cap of just under 36 million tons, 35% lower than what the states have agreed to. More ambitious electric sector decarbonization would make it possible for the region to achieve needed emission reductions through the electrification of transportation and other sectors.[37]

In brief summary, RGGI has done much good and can do even more.

California: Benefits and Lessons Learned

Moving now to the state level, California has one of the best known cap-and-trade programs. The state stands out on many counts: it has the largest population (39.5 million), the largest population of young people (almost 9 million); the largest economy ($2.6 trillion), and the largest cap-and-trade market of any state in the United States. California also stands out as having set new records for heat and acreage lost to wildfires. In 2006, the California legislature passed the Global Warming Solutions Act (AB 32) requiring the California Air Resources Board, known as CARB, to undertake a statewide effort to reduce global warming pollution. After extensive stakeholder input and with much research and analysis, CARB decided that cap-and-trade regulation should be one of the key measures used to cut GHG emissions. California kicked off its emissions trading program in 2013, with the goal of reducing the state's GHG emissions to the 1990 level by 2020. The program regulates GHG emissions economy-wide: from electricity generation, manufacturing, and cement production, to oil and gas production and supply. By some metrics, it has been a major success, but it has also been criticized for failing to deliver benefits to the communities that need them most. These criticisms have led to many changes in the program, which is still evolving. This history provides valuable lessons for others considering cap-and-trade or even other approaches to climate change mitigation.[41]

What have been the benefits? Since its initiation in 2013, overall emissions of GHGs and co-pollutants including fine particles from regulated entities have decreased despite substantial growth in the economy. While a good part of the credit goes to cap-and-trade, complementary measures to mitigate climate change, such as renewable portfolio standards for electric utilities, the zero-emission vehicle programs, and low carbon fuel standards for automobiles, have played a big role. As a result of these initiatives, in 2016, the state met its goal of rolling back carbon emissions to 1990 levels by 2020 and is in a strong position to reach its next target of reducing GHG emissions by an additional 40% below 1990 levels by 2030. Ultimately, California aims to reduce GHGs 80% below 1990 levels by 2050.[42]

As a result of the decreasing cap, emissions of GHGs from the regulated sources declined by more than 60 million tons between 2015 and 2020. To date, the cap-and-trade auctions have generated $14.9 billion, of which $14 billion has been allocated to programs that support reduced air pollution through community emissions reduction programs, incentives for cleaner vehicles and equipment, transit, and transit-oriented affordable housing projects. Cumulatively, these investments since 2015 are expected to reduce the emissions of GHGs (CO_2 equivalents) by at least 66 million metric tons over project lifetimes. Air pollutant emissions have also been cut: to date, the investments have reduced total emissions by an estimated 60,000 tons, including 37,500 tons of NO_x, 1,900 tons of diesel PM, 3,300 tons of $PM_{2.5}$, and 11,600 tons of reactive organic gases that form ozone. Not included are the reduced emissions resulting from investments of auction proceeds in the California High-Speed Rail Authority.[43]

The overall picture of health benefits from air pollution reduction was also positive. The cap-and-trade regulations, together with the low carbon fuel standard, have been projected to avoid 470 premature deaths, more than 14,000 cases of upper and lower respiratory symptoms, 239,000 cases of acute respiratory symptoms, 464 cases of acute bronchitis, 21,000 asthma attacks, and 338 cardiac and respiratory hospitalizations between

2010 and 2020. The estimated economic value of these avoided deaths and illnesses is $16.7 billion. Combined with $3 billion in benefits of GHG reduction, the total benefits of $19.7 billion are almost five times the costs of implementing the program. These estimated benefits are based on projections, therefore uncertain. They also largely pertain to adults, omitting the avoided cases of preterm birth, low birth weight, new cases of asthma, asthma attacks, and autism in children, which can have very large long-term costs. Nevertheless, viewed as a whole, cap-and-trade has been successful in addressing climate change and improving public health.[44]

Nevertheless, once again, success has not been unalloyed. Concerns about impacts on the least advantaged, "environmental justice" communities were voiced early on but were initially discounted. Environmental justice groups had been instrumental in shaping the language of the California's AB 32 that, among other things, created the state's Environmental Justice Advisory Committee. As the cap-and-trade program was being developed, the Committee expressed concern that the program would disproportionately harm communities of color and low-income neighborhoods. In his book *Climate Change from the Streets*, Michael Mendez, Assistant Professor of environmental planning and policy at the University of California Irvine, tells us that, since they did not produce analyses supporting their claims due to a lack of capacity and resources, their concern was dismissed as being led more by emotion than science.[45]

Scientific data arrived in the form of a peer-reviewed report by researchers at the University of California Berkeley that found that, during the first several years of the program (from 2013 to 2015), most regulated entities had increases in their annual average local GHG and $PM_{2.5}$ emissions. Neighborhoods that experienced increases rather than decreases in GHG and co-pollutant emissions had higher proportions of low-income residents and people of color. The lack of local air quality benefits was ascribed to the ability of companies to purchase offsets elsewhere and reduce their use of carbon-intensive energy generated out of state. CARB, which implements and enforces the cap-and-trade program, has labeled this and other reports of disproportionate impacts to be "inconclusive". However, in 2021, the Board reaffirmed its commitment to measures that directly reduce local pollution and reduce exposure for environmental justice community residents. It noted that, cumulatively, 50%, or just over $4.0 billion, in implemented California Climate Investment dollars has benefited priority populations.[46]

How was this achieved? Mendez recounts how, through conflict and collaboration, California's environmental justice movement became a key and sophisticated player in shaping cap-and-trade and climate policy in general. Through community engagement, political mobilizations, education, and community-based research in partnership with academic investigators, the movement transformed the global problem of climate change to one centered on public health at the local level. One such research collaboration developed a screening tool to systematically identify communities most burdened by multiple sources of pollution and most vulnerable due to health and socioeconomic disadvantage. In large part through the efforts of the environmental justice movement, by 2016, 35% of the revenues from cap and trade were required by law to be directed to environmentally and socially disadvantaged communities. Another milestone was the passage of legislation, in 2017, establishing a partnership between communities and regulators in creating legally binding roadmaps for reducing local air pollution in disadvantaged communities. Mendez recaps this evolution: "Through a continual practice of conflict and collaboration, environmental justice communities are no longer merely perceived as impacted; they are now official knowledge producers in climate change governance."[47]

So, we may declare the California cap-and-trade program a fair success, as much for the lessons it provides on the central role of environmental justice and community knowledge in climate policy as for the reductions in emissions and health costs that it has achieved.

New York City

New York City (NYC)'s 8.4 million residents face serious threats from fossil fuel air pollution and climate change, especially its 1.7 million young people. They are daily exposed to air pollution from transportation, buildings, and energy production, which are also the predominant sources of GHG emissions in the city. Although air quality in NYC has improved substantially over the past 10 years, the city has the distinction of having higher $PM_{2.5}$ concentrations than any other city in the US Northeast and Mid-Atlantic regions. The 50 most polluted census tracts in New York state are all in NYC (the boroughs of the Bronx, Manhattan, and Queens), where pollution is more than double the state average. An estimated 2,000 premature deaths and 6,500 emergency department visits and hospitalizations each year in NYC are attributable to $PM_{2.5}$ alone. However, the burden is inequitably distributed among the state's racial groups, with African American, Latino, and Asian American New Yorkers having higher exposure than White New Yorkers. In the West Bronx, the region's most polluted census tract, about 70% of the residents are Latino and 30% African American. Inequity in exposure relative to racial and ethnic groups is more pronounced than are disparities by income.[48]

NYC is also one of most vulnerable cities in the United States to climate change. In 2019, the NYC Panel on Climate Change reported that climate change is increasingly present in everyday New York as extreme weather events, high temperatures in summer, and heavy downpours increase, and sea level rises. The report found that areas with lower incomes and the highest percentages of African American and Hispanic residents are consistently more likely to suffer the impacts of climate change. Scientists have projected that more than 3,000 New Yorkers could suffer heat-related deaths every year by the 2080s unless action is taken to sharply reduce GHG emissions and put in place adaptation strategies. The city has suffered mounting economic losses from climate change. A single event, Hurricane Sandy, laid bare the city's vulnerability to climate change in 2012, inflicting more than $19 billion in damages.[49]

Over the past two decades, our research at the Columbia Center for Children's Environmental Health in partnership with West Harlem Environmental Action (WE ACT) has documented associations between air pollution and multiple health effects in the young, as well as interactions of air pollution with stress due to poverty, underscoring their combined role in health disparities (see Part I, Chapters 3 and 4). This research has also yielded some good news by revealing the health benefits of various policies that targeted specific known air pollution sources. For example, to tackle pollution from city buses, between 2001 and 2013 the NYC Metropolitan Transit Authority shifted the bus fleet to cleaner, low-sulfur diesel fuel and increased the number of hybrid electric buses. Our analysis, in collaboration with lead investigators at Drexel University, has found that these actions improved air quality near the most traveled bus routes.

To reduce emissions from taxis, in 2005 and 2006, the NYC Committee on Transportation passed several laws that required the Taxi and Limousine Commission (TLC) to approve hybrid vehicle models for use as taxis, required that 9% of new fleet medallions be sold exclusively to owners of hybrid or natural gas vehicles, and provided incentives for the purchase of taxi models classified as "clean air" vehicles by the US EPA. Using monitoring data from the NYC Community Air Survey and taxi inspection data from the TLC, we found that, between 2004 and 2015, taxi exhaust emissions of nitric oxide (a precursor to ozone) dropped by 82% and total PM fell by 49%. The fuel efficiency of the NYC taxi fleet more than doubled, from 15.7 to 33 miles per gallon. New York's clean air taxi legislation was clearly effective in boosting fuel efficiency and improving the air New Yorkers breathe.[50]

Another initiative aimed to reduce air pollution from residential buildings. In 2011, the NYC Clean Heat Rule required buildings to transition from use of the dirtiest residual diesel-based fuel (no. 6) to no. 4 (less dirty) by 2015, then, by 2030, to switch to the cleaner low-sulfur no. 2 oil or natural gas. This was important because burning of residual fuels was a major source of air pollution in NYC, producing higher levels of PM than any other diesel type, mainly because of diesel's high sulfur content. By November 2015, most buildings in the city had complied with the ban on no. 6 oil. The next step has proved more difficult: boiler upgrades to the cleaner no. 2 oil or natural gas are costly, so buildings in higher income neighborhoods have been more likely to switch directly from burning no. 6 or no. 4 oil to the cleaner fuels than buildings in lower income neighborhoods. In 2018, 1,724 (53%) of the buildings still burning residual fuel oil were situated in disadvantaged neighborhoods in Northern Manhattan and the Bronx. Overall, however, the program has been successful in reducing exposure to toxic air pollution that is clearly linked to health effects in children and adults. Our studies have found that for every 10 buildings that converted from heating oil no. 6 to cleaner fuels, there were significant reductions in monitored levels of sulfur dioxide, $PM_{2.5}$, and NO_2.[51]

Turning to the health benefits of the complete phase-out of residual oil mandated by 2030, the reduction in annual average $PM_{2.5}$ would avoid an estimated 290 premature deaths, 180 hospital admissions for respiratory and cardiovascular disease, and 550 emergency department visits for asthma every year. The largest improvements would be seen in areas of highest building and population density.[52]

We also have had the opportunity to assess the trend in air pollution exposure among women who were enrolled into our NYC cohort studies during the time period when these and other measures were being implemented to reduce emissions from fossil fuel burning. Among 700 women monitored during pregnancy for levels of exposure to airborne PAHs, we found a highly significant downward trend in PAH exposure between 1998 and 2019.

Through a combination of city- and cohort-level analyses, we have been able to demonstrate that policies to control fossil fuel combustion emissions from various sources in the city *did* benefit New Yorkers, including pregnant women and their children. Since NYC is in no way unique in having a large inner-population at risk from air pollution, these results are relevant to policymakers in other cities.

Clean Air in NYC During the COVID Shutdown: A Goal for Policymakers

Because good news from many sources is needed to encourage and incentivize the public and policymakers, we come to a "natural experiment" in clean air during the NYC COVID-19 shutdown. The tragedy of the COVID-19 epidemic gave us a unique opportunity to assess the health benefits to NYC children from dramatically cleaner air as seen during the COVID-19 shutdown (March 15, 2020–May 15, 2020). $PM_{2.5}$ concentrations during the shutdown were about 23% lower than the average of the corresponding weeks in 2015–2018. We calculated that, if this level of improvement were sustained for the next 5 years, the health co-benefits in terms of avoided cases would be substantial: they included 1,200 preterm and term low-birth-weight babies, more than 5,700 emergency department visits and hospitalizations in asthmatic children, almost 2,000 new cases of child asthma, hundreds of cases of autism, and as many as 7,800 cases of adult and infant mortality. The total city-wide savings over this period would be as high as $77 billion. Although the drop in air pollution during the shutdown resulted from efforts to control the epidemic, the preceding case studies have shown us that a comparable improvement in air quality could be achieved by policies that have been implemented around the world to reduce fossil fuel emissions.[53]

CONCLUSION

These stories bring air pollution and climate solutions out of the realm of abstraction. They tell us not only that change is possible but also that it has been achieved in real settings around the world, with very substantial benefits to public health—particularly the health of our children. In cases where cost-benefit analyses have been done, the benefits clearly outweigh the costs. As we saw with the London case study and California climate policy, these interventions are not static but continually evolve based on experience and input from environmental justice groups and other stakeholders. We can refer to them as success stories for all these reasons. They equip us to tackle the final topic: How do we—as a society, as policymakers, as citizens, as individuals, as parents and grandparents—take action now to avoid climate catastrophe for our children?

CHAPTER 7
Solutions Now

INTRODUCTION

We have seen up close the vulnerability of fetuses, infants, young children, and adolescents to air pollution and climate change from fossil fuel burning and the myriad physical and mental health effects being inflicted on them now. By examining the threats of fossil fuel holistically—moving them out of their traditional silos—we now understand that we are facing a public health emergency for our children. We know that all children are being harmed, but we are shocked by the greater suffering of poor children and children of color. We have been cheered by the health benefits to children from policies and other initiatives that reduce exposure to fossil fuel pollution. We have read fact after fact detailing these realities. Now we must convert that knowledge to the wisdom to take action.[1]

What action is needed to protect the health and well-being of children from fossil fuel pollution and climate change, from the time they are conceived, throughout their lives and the lives of the next generations? What is needed to ensure that the solutions are applied so that all children are treated equally? The answer is that we must act simultaneously on two fronts: to protect children today from climate hazards (adaptation) and to attack the root problem by dramatically reducing greenhouse gas (GHG) emissions and strengthening natural carbon sinks (mitigation). We must ensure that the most socially and economically vulnerable children are protected. Although the focus in this chapter is on mitigation of climate change through emissions reductions, it is clear that immediate action is needed to help children adapt to the climate and environment of "now" by limiting their exposure to climate disasters while at the same time reducing

Children's Health and the Peril of Climate Change. Frederica Perera, Oxford University Press. © Oxford University Press 2022. DOI: 10.1093/oso/9780197588161.003.0007

their vulnerabilities. The United Nations Children's Fund has reported that, right now, almost every child on Earth is exposed to at least one climate and environmental hazard, shock, or stress; approximately one-third of all children are exposed to four or more stresses; and half the world's children (1 billion) live in countries that are at an extremely high risk from the impacts of climate change. Their conclusion: we must invest heavily in adaptation and disaster risk reduction for the children now alive and the 4.2 billion children who will be born over the next 30 years "or they will face increasingly high risks to their survival and well-being."[2]

Protecting children who now face water scarcity, coastal and river flooding, heatwaves, cyclones, pollution, and climate-related infectious disease will take coordinated efforts on the part of national and local governments, communities, schools, and businesses. Immediate solutions include providing clean water to children and families facing drought and water contamination, installation of early warning systems for flooding and air pollution, and training and evacuation planning for families and children. To protect children from heatwaves, shaded areas must be provided where they play, live, and learn. Providing mosquito nets will help protect children from malaria and dengue which are broadening their reach due to climate change. These more immediate climate-specific measures must be paired with broad social and public health programs to reduce poverty, respond quickly to crises, and provide water, sanitation, and hygiene services, high quality healthcare, nutritional support, and education. Teaching children about climate hazards and what they can do to protect themselves will empower them as agents of change.[3]

Turning now to climate change mitigation, international expert groups have not directly asked "What must we do to save the children?" Instead, they have posed the question: "How can we act in time to avoid disastrous warming, holding it to no more than 1.5°C (2.7°F) above preindustrial levels?" Their response has been to echo the words of the International Panel on Climate Change (IPCC): we must slash global carbon emissions by 45% from 2010 levels by 2030 and reach net zero carbon around 2050. At the same time, they say, we must strengthen and support natural carbon sinks that absorb more carbon dioxide (CO_2) from the atmosphere than they release.

A pause here for definitions:

> *Net zero CO_2 emissions* are achieved when anthropogenic CO_2 emissions are balanced globally by anthropogenic CO_2 removals (natural carbon sinks or technology) over a specified period.

Zero emissions would result from setting anthropogenic emissions to zero.

Drawdown is the point when levels of GHGs in the atmosphere stop climbing and start to steadily decline.

Youth and other activists object to the distant goal of net zero by 2050. Greta Thunberg and other school climate strikers argue that distant net zero emission targets will mean absolutely nothing if present high emissions continue even for a few years. We don't just need goals for 2030 or 2050, they say, but we need them for every month and year henceforth. "And until we have the technologies that at scale can put our emissions to minus then we must forget about net zero or 'carbon neutrality'. We need real zero." The young activists—exasperated—have exhorted us adults "to go home, study the facts and come back when you've done your homework."[4]

Expert groups have done our homework for us and all agree that we have the tools now to drastically reduce emissions and support carbon sinks, if we scale them and apply them as a system. No technological leaps are needed. This is welcome news to many of us who have wrung our hands over the enormity of the problem and decided it was hopeless to try to tackle it.

All agree that we can and must meet the urgency of the situation with aggressive action. We hear again the words of the young poet, Amanda Gorman:

There is no rehearsal. The time is
Now
Now
Now.[5]

AVAILABLE SOLUTIONS

Viewing the available solutions through the lens of the urgent needs of children, we can group them into four phases or waves as outlined by the nonprofit research group Project Drawdown: those that can sharply and rapidly reduce emissions and bolster natural sinks ("quick wins"), those that can make steady progress over the next years (infrastructure improvement), others that are naturally slower (growing natural sinks), and those requiring decades of research and development (deploying new technologies) (Figure 7.1). Concurrent with all these phases are the social and policy changes to redress inequality in the health impacts of fossil fuel.[6]

There is no silver bullet. We need all these solutions, pursued in parallel and woven together over time, to make the most progress possible, year by year, decade by decade. Some of the solutions in this chapter are really only available in the short-term to the developed countries; others, like solar panels, green and cool roofs, and clean cookstoves are available in developing countries. All are integral to sustainable development. Although the solutions are framed as the means of mitigating climate change, in reality many of the methods and technologies that curb climate change also cope with its impacts and help communities adapt and become more resilient. There are certainly obstacles to implementing all of these solutions, but they offer large emissions reductions plus strong economic returns. The costs are minimal compared to the health co-benefits, the savings that outweigh costs four to five times over, and the opportunity to secure a viable future for our children and generations to come.[7]

All experts agree that the solutions must be used in ways that advance social and economic equity. This is not just a question of compassion; it is one of efficiency. It turns out that climate policymaking can deliver simultaneously on social and climate goals. A group of experts carried out a systematic review of more than 200 climate policies that have been implemented around the world, albeit mostly in Western, industrialized, and democratic countries. The policies ranged from taxes to development

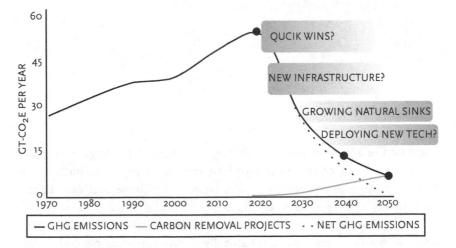

Figure 7.1 A hypothetical example of how four different "waves" of climate solutions might unfold over time, depending on our choices.
Credit: Jonathan Foley © 2021.

of renewable energy sources, subsidies, direct procurement, and large renewable deployment projects. The group found many examples across different countries and policy types that delivered on both social and climate mitigation goals. Keys to success, they concluded, were policy designs that attended to basic issues of equity and procedural justice; provided scope for citizen, community, and public participation in decision-making; and were supported by well-functioning institutions. The concept that climate policy, properly designed, can be a tool for social justice may be new to many of us.[8]

It is also encouraging that many of the climate solutions aimed at reducing sources of emissions—such as reducing waste, changing diets, protecting and restoring natural sinks, and shifting agricultural practices to reduce emissions—boost carbon sequestration at the same time. The social supports required for adaptation also overlap with those that support climate change mitigation.[9]

Quick Wins

The first steps in protecting children's health and securing a viable future for them are to immediately stop the new production of coal, oil, and gas; halt development of projects that increase carbon emissions; and rapidly shift electricity production from coal and other fossil fuels to wind turbines, solar power, and geothermal power. The transition has begun: as mentioned in Chapter 1, global coal consumption has leveled off since 2010 and is dropping in many parts of the world. Growth in global renewable energy capacity in 2020 beat all previous records: more than 80% of all new electricity capacity added in 2020 was renewable, with solar and wind accounting for 91% of renewable growth: 2021 set another annual record, driven by solar photovoltaics. Renewable electricity capacity is expected to increase by more than 60% in the next 5 years, reaching a capacity equivalent to that of fossil fuels and nuclear today.[10]

Additional rapid returns can come from increasing efficiency in electricity use (especially in buildings and industrial processes) and shifting to energy-efficient transportation. Buildings use more than half of all electricity, so boosting their energy efficiency through retrofitting or design gives a quick win. The technologies already exist and include LED lighting; automation systems to control heating, cooling, lighting, and appliances in commercial buildings; district heating rather than on-site systems; insulation; smart thermostats; solar panels; high-efficiency heat pumps; green roofs using vegetation as living insulation; and white, cool roofs to reflect the sun's energy. They can mean large energy savings: in the United States, for example, by 2027, widespread LED use could save the equivalent annual electrical output of 44 large electric power plants with a total savings of more than $30 billion.[11]

A community-initiated program is demonstrating how low- and middle-income communities and communities of color can successfully make a transition to affordable, renewable energy for buildings while also creating green jobs for local residents. In 2016, the environmental justice group, West Harlem Environmental Action (WE ACT), and its partners launched Solar Uptown Now, a campaign to bring Northern Manhattan community members together to purchase solar panels as a group, bringing down the cost of installation in their buildings. This partnership has resulted in the installation of at least 415 kilowatts of new solar power on affordable housing units in Northern Manhattan. Bringing this type of program to scale will reduce pollution from local power plants and result in cleaner air, thus benefiting the health of children and other residents. The program has provided solar installation training to more than 100 unemployed and underemployed Northern Manhattan residents, and some have already landed jobs in the solar power industry.[12]

With respect to industry, the dominant mode of industrial operation has been characterized as "take-make-use-trash," an unsustainable linear flow of materials. Industry operations at the plant or factory are directly responsible for 21% of GHG emissions, plus nearly 50% of off-site electricity generation emissions and vast amounts of waste. The biggest emitters are manufacturers of cement, iron, and steel, followed by producers of aluminum, fertilizer, paper, plastics, processed foods, and textiles. A quick-win solution is to enhance energy efficiency in these industrial processes and redesign the processes themselves to lower GHG emissions.[13]

We may be surprised to learn that reducing industrial production of hydrofluorocarbons (known as HFCs) has a huge impact. HFCs are primarily manufactured for use as refrigerants and coolants, replacing ozone-depleting substances being phased out under the Montreal Protocol to protect the ozone layer. But some HFCs have a global warming potential more than 3,000 times that of CO_2 over a 20-year period and are the fastest growing GHGs. Available solutions include better management of leaks and disposal of HFCs, and their substitution with ammonia, captured CO_2, water vapor, and hydrocarbons. In 2021, the United States began to phase-out the use of climate-warming HFCs, joining many other countries in this effort.

In addition to curbing emissions of GHGs, providing a sustainable future for our children means moving toward a circular economy as rapidly as possible. A critical step, beyond waste prevention, is to put to use the massive amounts of waste from industrial processes and materials production. The technologies are available now at the scale needed to recycle and reuse materials, compost organic material, and convert waste to heat or electricity. For example, methane from landfills is being captured and converted to a nutrient-rich fertilizer and biogas.

Transport accounts for 24% of global direct CO_2 emissions from fuel combustion (29% in the United States). Cars, trucks, buses, and two- and three- wheelers account for nearly three quarters of CO_2 emissions from transport; aviation and shipping supply the rest. Various alternatives to using gasoline and diesel fuel for transportation are being applied around the world. The highest impact solution is the electrification of cars, buses, and trucks paired with renewable electricity generation. Other ready solutions are expansion of public transit, increasing walkability, and providing bicycle infrastructure.[14]

More and more cities around the world are increasing the opportunities for people to walk or bike instead of ride. The world's most walkable cities include London, Paris, Bogotá, and Hong Kong according to a report by the Institute for Transportation and Development Policy. The cities were ranked according to proximity to car-free spaces, schools and healthcare services, and the overall shortness of journeys. Big cities in the United States scored low by these criteria. Benefits of walkability include fewer traffic fatalities, less obesity and chronic disease, improvement in mental health and happiness, greater social cohesion, and equality. Walking to school for children fosters independence, prevents obesity, and can even improve academic performance. In Copenhagen, 40% of schoolchildren now walk to school.[15]

The smartest cities have also encouraged bicycling. Since the 1960s, Copenhagen has restructured the street network to better serve pedestrians and cyclists, reclaiming space from vehicle travel lanes and on-street parking to create active networks. Copenhagen's main street, Strøget, increased pedestrian use by 35% in the first year of the program alone. Copenhagen has seen cycling traffic become a primary mode of transportation, with 41% of adults biking to work or school and 25% of children biking to and from school. In Odense, Denmark, a city of around 200,000 people, most adults bike, and four out of five children bike, walk, or skateboard to school.[16]

Other quick wins are action on the short-lived but powerful GHGs: methane and black carbon. Given that both these pollutants have an immediate effect on global warming, whereas CO_2 warming is more gradual, cutting these emissions now gives us time to tackle other more intractable sources. For example, methane has a short lifetime (about 10 years) but has more than 80 times the global warming impact as CO_2 in the near term. The immediate solutions are to end the practice of venting or "flaring" during the process of oil and gas development and plug methane leaks from natural gas wells and pipelines. There has been progress on the policy front. In September 2021, the United States and Europe pledged to work to cut global methane emissions by a third in the coming decade and urged other nations to join their effort.[17]

We hear less about black carbon emitted from dirty cookstoves, biomass burning, and other combustion sources. Even more short-lived than methane, black carbon is nonetheless a plentiful and potent GHG. Here, too, there has been progress. Organizations like the World Bank and the Clean Cooking Alliance have made headway in providing clean cooking fuels and technologies to the 2.8 billion people in the world lacking access. A rapid transition to clean cookstoves in developing countries will benefit both the climate and the health of women and children in developing countries. In fact, awareness of the health impacts have been the major driver of efforts to curb these emissions.

We owe our awareness of the health effects of cookstove emissions to the pioneering work of the late Kirk Smith, a professor of global environmental health at the University of California Berkeley School of Public Health and director of the Collaborative Clean Air Policy Center in Delhi, India. His research in developing countries showed that the daily task of cooking with biomass fuels posed a substantial risk to the health of women and children in low-income populations. His work pushed policymakers and the private sector to recognize that providing cleaner fuels such as liquefied petroleum gas and electricity was a matter not only of public health and a safer climate but of equity and fairness. I met with Kirk on several occasions and each time was inspired by his commitment to helping vulnerable women and children.[18]

Finally, drastically cutting food waste and shifting from diets high on the food chain will have immediate payoffs for the climate and health. Most of us are not aware that roughly a third of food produced is wasted. The solutions include simple low-cost food storage methods such as small metal storage silos for small-scale farmers in the developing world, food redistribution programs, better labeling of end dates on products so food is not unnecessarily thrown away, and reduction in portion sizes. Diets high on the food chain are major sources of GHGs: ready solutions are lower consumption of meat and dairy and more plant-rich meals.[19]

Here the picture is mixed. Global meat consumption is on the rise, with the fastest growth in low- and middle-income countries. However, according to a recent survey, 40% of Europeans in 11 countries have reduced their meat intake or stopped altogether. In the United States, public health campaigns have driven down beef consumption per capita; however, consumption is still higher than in almost any other country. Among the available solutions are educational and other programs such as "meatless Mondays" to promote dietary changes in young and old, alternative meats and proteins, and eco-friendly livestock management.[20]

The term "quick wins" glosses over the reality that these solutions not only require immediate changes in government policy, industrial operations, and the design of cities, but also in individual behaviors. These in turn require cultural shifts and the building of political power, as we will see. Although these are all major challenges, the health and economic gains more than offset the costs, and children will benefit the most.

Steady Progress: New Infrastructure

In addition to shutting down fossil fuel energy sources and deploying renewable energy across the planet as quickly as possible, we simultaneously need to put in place the infrastructure required to convert the heating and cooling systems of buildings to renewable electricity and to replace gas and diesel powered vehicles with electric bikes, cars, trucks, and trains. Completion of this energy transition will inevitably require the 2020s and 2030s even under the most aggressive scenarios.[21]

A stumbling block has been the lack of adequate infrastructure to transmit and store electricity from renewables, because renewables are more variable than conventional electricity generation, a flexible system is needed. This includes *microgrids*, localized groupings of electricity generation technologies, paired with energy storage or backup generation to manage demand. Energy storage can be done through dispersed, standalone batteries or large utility-scale batteries, which are able to store excess energy when solar panels and wind farms are producing electricity and deliver it back into the grid when they are not. As a result of steeply falling prices and technological progress, grid-scale storage systems are seeing rapid growth. In 2017, Australia was the first country to install a major storage battery on its grid. In California, a large lithium-ion battery is already online in San Diego, and others are in various stages of completion or development. Plans have been announced for super-sized batteries in South Florida, the United Kingdom, Lithuania, Germany, Chile, and Saudi Arabia. There is a race to drive down the cost of long-duration storage. For example, as part of its Energy Earthshots Initiative, the US Energy Department is directing resources to improving technologies and cutting costs by 90%. In November 2021, the US Congress passed a bill committing major funds to upgrading the power infrastructure, including clean energy transmission lines and a flexible and resilient electric grid.[22]

Shifts in agricultural practices from high-emitting techniques to low-emissions methods, including so-called regenerative practices, will make steady progress in bringing down emissions of GHGs from cropland, pastures, and rice fields. Agricultural production currently accounts for about 10% of GHG emissions in the United States; the percent is far larger globally. Specific solutions include avoiding the overuse of nitrogen fertilizers that create nitrous oxide (a GHG) and improvement in rice production by alternately wetting and drying rather than flooding rice fields to minimize methane emissions. Conservation techniques—using cover crops, crop rotation, and minimal tilling—avoids CO_2 emissions and sequesters carbon. Regenerative agriculture based on compost application, green manure, and organic production further reduces GHG emissions. All are proven techniques. Organizations all around the world are helping farmers transition from conventional to regenerative agricultural practices. A central element of the European Green Deal is a shift to a more sustainable system of agricultural production that includes regenerative farming.[23]

Benefits go far beyond sustainable food production to preservation of community and culture. Wendell Berry, a well-known Kentuckian farmer, essayist, and poet, has long advocated regenerative agriculture, arguing that today's industrial agricultural practices are harmful to ecology, community, and the local economies that the farms have traditionally served. Since publication of his 1977 book, *The Unsettling of America*, Berry has made the compelling case that responsible, small-scale agriculture is essential to the preservation of the land and the culture. He has taken on agricultural policy in the United States, calling for a farm policy that would rejuvenate and regenerate our nation's ecosystems. Such a policy shift still awaits, but recent trends in the global food sector suggest that regenerative agriculture will become more mainstream: consumer demand is increasing, companies are seeing the value of sustainability credentials, and the public is becoming more aware of the need to tackle climate change.[24]

Benefits to children from these infrastructure changes and a major shift in food production are better health from a safer climate, lower air pollution, food free of pesticides, and the preservation of the biodiversity and ecosystems on which their future depends. Improving agriculture is a climate solution that can increase family income and lead to better education for children. These changes take time, so it is heartening that the transitions have already begun.

Protecting, Growing, and Restoring Natural Sinks

Protection and restoration of nature's carbon sinks is imperative if we are to protect children and their future. So, too, is the creation of new ones. This wave of climate solutions is, by its very nature, gradual, so though we should (and must) start today, our children won't see the full effects of these solutions for some time.

Halting destruction of ecosystems provides dual benefits: preventing the release of carbon from vegetation and soil and supporting their ability to sequester carbon. Top among natural emitters of carbon are peatlands and forests. Peatlands are a type of wetlands that occur in almost every country on Earth, from bog landscapes in Scotland to swamp forests in Southeast Asia and tropical mountain peatlands in South America. Drainage, agricultural conversion, and mining of these vulnerable lands for fuel have taken a severe toll so that damaged peatlands now contribute about 10% of GHG emissions from the land use sector. Solutions include peatlands protection and rewetting to reduce emissions and allow continuing carbon sequestration.[25]

As we read in the first chapter, forests covering 31% of the land area on Earth play a critical role in accumulating and storing CO_2 that otherwise would contribute to climate change. Halting destruction of forests is therefore essential to preventing the release of the vast quantities of carbon stored in their biomass and soil. Forests are being lost at an alarming rate. About half of the breakneck deforestation has occurred in the Amazon, the largest tropical forest on the planet, where the pace of deforestation is now at its highest rate in a decade. The Amazon is on a path to become a net emitter of GHGs. Solutions are reform of land use policy, enforcement of existing laws in Brazil, securing Indigenous land tenure, and funding by other countries in return for protection of the Amazon.

Securing Indigenous peoples' forest tenure not only safeguards their rights but also ensures the continuation of their traditional practices that protect ecosystems and prevent emissions from deforestation. Protecting tropical forests, for example, can provide 30% or more of the GHG emission abatement needed to halt climate change. In Brazil, about 2 million hectares of Indigenous land are still awaiting official land tenure confirmation, so the Indigenous people there do not have their full property rights. This is a missed opportunity: an analysis comparing Indigenous territories with and without land tenure in the Amazon found that secure property rights significantly reduced the levels of deforestation inside Indigenous territories—by orders of magnitude. A World Resources Institute study calculated that secure forest tenure for Indigenous people in Bolivia, Brazil, and Colombia could reduce emissions by 43–60 million metric tons of CO_2 equivalent through 2035, with estimated benefits from $700 to $1,500 billion.[26]

Coastal and ocean sinks play a massive role in sequestering carbon. Since the 1980s, oceans have absorbed 20–30% of human-created CO_2 but their capacity to soak up excess carbon is running out. The solutions include protecting and restoring the "blue carbon" ecosystems—mangroves, salt marshes, and seagrass meadows (collectively, *coastal wetlands*)—that are essential in supporting photosynthesis and carbon storage. Coastal wetlands not only absorb energy from storm surges and help limit shoreline erosion, but they also filter water and serve as nursery grounds for a large number of species. Their protection is therefore a "triple win" in addressing climate change.[27]

Take mangroves, for example. While they cover only 0.1% of Earth's land surface, they store more carbon per hectare than any other type of forest. Half of the world's mangroves have been lost in the past 50 years as a result of rising sea levels, oil spills, building of coastal infrastructure, and urban expansion. Because so many mangroves have been degraded, it is critical to replenish and revive these ecosystems. To that end, a coalition of nonprofits (the World Wildlife Fund, Conservation International, the International Union for Conservation of Nature, the Nature Conservancy, and Wetlands International) have formed the Global Mangrove Alliance. Their goal is to partner with communities, other stakeholders, and governments around the world to increase mangrove habitat by 20% by 2030. They have been working in countries as diverse as Papua New Guinea, Indonesia, Malaysia, Sumatra, Pakistan, Madagascar, Mozambique, Tanzania, Guinea Bissau, Senegal, and Colombia using a holistic approach to help mangrove forests replenish and flourish on their own with minimal human-led replanting (Figure 7.2).[28]

A positive note: more than 50 countries have recently pledged to work toward preserving 30% of the world's lands and oceans in their natural or near-natural state by 2030.[29]

New Technologies

The fourth wave is the development and deploying of engineered sinks that require decades of research and development. For example, a variety

Figure 7.2 Mangrove restoration efforts can help reverse declines of these valuable trees. That was the goal of this seedling planting in abandoned shrimp ponds near Jaring Halus Village, North Sumatra.
Credit: Tim Laman/Nature Picture Library/Minden Pictures.

of industrial and chemical processes are being explored to pull carbon from the exhaust gases of a power plant or industrial process (*carbon capture*) or from the air. However, it would be a big mistake to let the promise of carbon removal technology be an excuse to delay immediate emissions cuts. These technologies are likely decades away and may only be able to absorb a portion of our emissions.[30]

An example of a developing technology is the new carbon capture and storage plant in Iceland called Orca, intended to capture 4,000 metric tons of CO_2 per year and bury it underground. Orca has been described as the start of a complex, multidecade, global deployment process that follows years of research and development. Other ideas are to use new technologies to make carbon-free cement and fossil-free steel, plastic, and jet fuel. However, they are now far from economically competitive and scalable and are considered "coming attractions."[31]

Hydrogen is being touted as a clean fuel that, when consumed in a fuel cell, produces only water as a by-product. However, the source of the energy used to produce hydrogen is key to its usefulness in mitigating climate change. "Green hydrogen" refers to fuel created using entirely renewable energy sources, but today only 1% of hydrogen fuel is produced this way. The other 99% is extracted from fossil fuels, most from natural gas in a process that is energy-intensive and emits large amounts of CO_2 and methane. Replacing fossil-fuel–based hydrogen with green hydrogen is a priority.[32]

There is another priority area of climate change mitigation: the petrochemical industry. Industrial chemistry's use of oil is responsible for 14% of GHG emissions. Fossil fuels are both the feedstock and the source of energy used to make petrochemicals that are then transformed into hundreds of everyday products including plastics. One solution is to revamp the world's plastic system to reuse and recycle plastic in a circular economy. Another is *green chemistry*, a field that has been developing rapidly since the early 1990s. As the name implies, its goal is to design chemical processes using non-toxic and renewable green reactants, solvents, catalysts, and additives to synthesize environmentally friendly products. Although green chemistry is still a small part of the global chemical industry, it is seen as a key to sustainable development.[33]

CULTURE, POLICY, POWER, AND RIGHTS OF CHILDREN

From the above, we know that we have the knowledge and tools at hand to reduce emissions and strengthen children's resilience, but equally essential are the societal and political changes that enable and accelerate those solutions in an equitable way. This may be our greatest challenge, for these require nothing short of major cultural shifts that lead to the needed political and behavioral changes. Progress also requires laws and policies to address racism and socioeconomic inequality, elimination of those policies that have supported fossil fuel, and empowerment of environmental justice groups, children, and youth who have hitherto not had a say in their future. These requirements cannot be an afterthought, but are fundamental to meeting the challenge of climate change.

Cultural norms dictate what behaviors society considers acceptable or unacceptable, to be rewarded or discouraged. All the climate solutions described above involve changes in norms and behaviors at different levels: government, corporate, community, and individual. Cultural norms and the behaviors they dictate are not static, but evolving, driven by messages from the arts, media, and campaigns that reshape cultural narratives. In Chapter 5, "Power and Voice," we saw how the various players have worked to change harmful cultural norms and practices that have stymied progress on climate change. Through their testimony and actions, environmental justice groups, environmental organizations, Indigenous peoples, youth, and religious groups have helped to change values and norms. So, too, have poets and artists.[34]

The artist Maya Lin, American architect and sculptor, became famous even as a student for her design of the Vietnam Veterans Memorial in Washington, DC. Her art dramatizing environmental degradation and climate change has opened eyes and minds, most recently in an installation of dead trees in a Manhattan park. "Ghost Forest" was a towering stand of 49 dead Atlantic white cedar trees, which symbolized the devastation of climate change. But the dead trees were juxtaposed with the living trees in the park, symbolizing the possibility of regeneration. In fact, the trees are now being reborn as a boat that is being built by a group of teenagers in the Bronx, an underserved community in New York City. In the summer, the teenagers plan to sail the boat out on the Bronx River.[35]

The art of children has helped to engage young people and change public understanding of the urgency of climate change. It is also a poignant plea for adults to show leadership on climate change. The European Space Agency Kids hosts art competitions that look at different themes relating to the planet. Figure 7.3 shows the painting by a 7-year-old boy from India that was submitted to the Agency's climate change competition in 2018.[36]

Meeting the challenge of climate change requires a multifaceted effort to tackle social injustice, income inequality, and racial discrimination. Among the social policies needed to both help disadvantaged children adapt to climate change and accelerate mitigation are those providing quality education, access to mental and physical health services, reproductive healthcare for girls and young women, food, housing, and economic support. Additional accelerators of mitigation are policies to end de facto residential and school segregation. Echoing the proponents of the Green New Deal in the United States, in July 2020, Secretary-General of the UN, António Guterres, called for a new Global Deal "to ensure that power, wealth and opportunities are shared more broadly and fairly at the international level." He stressed the need to take account of the rights and dignity of every human being and the rights of future generations.[37]

To take just one example, most advanced, industrialized countries have child allowances, but the United States has been a laggard. Yet it has been estimated that increasing economic support to poor families and children by expanding the current US Child Tax Credit will return eight times the cost in improved health, educational attainment, future earnings of children, and both children's and parents' longevity. All children would benefit: children across the income distribution would see income gains.[38]

Ambitious international and national government policies on climate change are essential to cut global emissions at least by 45% by 2030, which

Figure 7.3 The European Space Agency Climate Change Competition—Earth—Rinny Joy.
Credit: John P. Anson, ESA Kids Climate Change Art Competition Winner, 2018.

is necessary to reach carbon neutrality by 2050. On a positive note, the 27-nation European Union, the United Kingdom, and the United States have recently announced more ambitious goals for reducing GHG emissions, but other big emitters—China, India, and Saudi Arabia—have not stepped up. And all countries need to do much more. A new analysis by the UN warned that, even if all countries carry through on the promises they made in Paris in 2015, the global average temperature will rise by at least 2.7°C (5°F) above preindustrial levels by century's end.[39]

States, regions, and cities are acting to reduce GHG emissions, as we saw in the case studies on the London congestion zone policy, the Regional GHG Initiative, and the California Climate Policy. In the United States, more than 20 states have set targets to achieve carbon-free electricity by 2050, and 130 cities are committed to achieve net-zero emissions in the 2040s or sooner.[40]

Reform of government policies that have historically favored fossil fuel is essential to reaching these goals. The United States and other countries have long provided subsidies to the fossil fuel industry as a means of encouraging domestic energy production. These are in the form of direct subsidies to corporations and tax benefits to the fossil fuel industry. Conservative estimates put US direct subsidies to the fossil fuel industry at about $20 billion per year and EU subsidies at €55 billion annually. Removing those subsidies and increasing incentives for renewable energy will level the playing field in which powerful fossil fuel interests have been so greatly favored up to now.[41]

So will divestment from fossil fuel, which is advocated by the UN, the World Bank, and many religious leaders. For example, issuing guidance on the implementation of his climate encyclical, in June 2020, Pope Francis called on Catholics around the world to divest themselves of fossil fuel companies. Divestment has now become a $14.5 trillion movement with over a thousand major investors, pension plans, and endowments committed. This is good for the climate and is also seen as a savvy business decision. In December 2020, New York State announced it will divest its $225 billion Common Retirement Fund from fossil fuels, decarbonizing the pension fund portfolio by 2040. This marked one of the largest pension fund divestments ever undertaken and followed years of advocacy led by the DivestNY coalition whose members include dozens of local and statewide organizations such as the Natural Resources Defense Council, New York Youth Climate Leaders, and the New York State Council of Churches. After years of lobbying and protests by students and alumni (including myself), in September 2021, Harvard University announced that it will divest its almost $42 billion endowment, the largest of any university in the world. Multilateral financial institutions have moved as well. In November 2019, the European Investment Bank agreed to end its multibillion-euro financing of oil and gas projects after 2021. The World Bank has made a similar move. At the most recent Climate Conference in Glasgow (COP26), 20 countries pledged to stop financing international fossil fuel projects—including oil and gas projects—by the end of 2022.[42]

Businesses have been major contributors to climate change. In fact, 100 fossil fuel–producing companies have been the source of more than 70% of global industrial GHG emissions since 1988 and more than 50% of those emissions since the Industrial Revolution. The major corporations that produce the vast number of products we buy, use, and throw away also bear heavy responsibility. There has been progress: more than 2,000 businesses around the world are working with the Science Based Targets Initiative that provides technical support to help them set emissions reduction targets in line with the goals of the Paris Agreement. And more than 300 businesses are committing to 100% renewable energy through another global initiative. However, only a third of 500 large companies listed on US stock exchanges have set ambitious targets for reducing GHG emissions, and they have had trouble matching deeds to words. Other big businesses have only weak goals. This has led to a call for requirements and standards, including from some corporate executives.[43]

In a hopeful sign, many businesses are beginning to move away from petrochemical-based products, with benefits for environmental health and climate change. Many thousands of widely used chemicals and their products are based on petrochemicals derived from oil and gas. Nonprofit groups like Clean Production Action in the United States have prompted market shifts away from toxic chemicals and plastics by providing companies with information on environmentally friendly alternatives. Clean Production Action has provided an accessible tool for companies to track their chemical footprints and measure their progress toward eliminating chemicals of high concern. Another nonprofit, Toxic-Free Future, has responded to the alarming upward trend in growth of the petrochemical industry—and its clear implications for climate change—by targeting its advocacy to the elimination of hazardous chemicals in plastics. The European-based International Chemical Secretariat (ChemSec) has a growing presence in the United States as well as Europe, with an online platform that offers leaders in safer production a means of differentiating themselves from laggard peers and gives innovators exposure to potential customers and investors. These are just a few of the groups that are working to shift business culture.

If government were designed to be truly responsive to people's needs, especially those of families with children, climate solutions would be more widely embraced. Reform of campaign finance laws, preservation of democracy, and protection of voting rights have become critically important tools for changing the power structure and making the government more responsive to what people need. Empowerment of environmental justice, community, and Indigenous groups and women, children, and youth who have valuable lived experiences and are most affected by climate change will benefit all of society. In Chapter 5, we saw the many diverse groups speaking out to build awareness and political support for action on climate change. Environmental justice and community groups have become increasingly powerful opponents of entrenched interests in industry and government. Chapter 6, "Success Stories," recounted how the environmental justice movement became a key player in shaping cap-and-trade and climate policy in California through community engagement, political mobilization, education, and community-based research. Leadership by community-based environmental justice groups like WE ACT and new coalitions between environmental justice groups and the "Big Greens" have created synergy in power building. The 2020 vote in the United States that relaced climate denier Donald Trump with Joseph Biden, whose major campaign promise was to address climate change and the economy, was an illustration of built political power.

Ensuring the fundamental rights of women and young girls to education and access to reproductive healthcare brings them greater political, social, and economic empowerment, and typically reduces fertility rates and population growth. Given the predicted surge in world population from 7.7 billion people today to as many as 10 billion in 2050 and the huge impact that such an enormous growth would have on energy consumption and waste, their empowerment is vital to climate solutions.[44]

Empowerment of children and youth allows them to bring their needs and experience to bear on the course of climate change mitigation and even to create and lead interventions such as tree planting, water testing, and planning what to do in a climate emergency. Young people argue, justifiably, that their futures are being destroyed, their rights violated, and their pleas ignored by governments. The UN Children's Fund states that the climate crisis is a child rights crisis, infringing on the rights of children enshrined in the Convention of the Rights of the Child to which all 196 eligible parties have signed, except the United States. In addition, the Paris Agreement calls on parties to respect and promote their obligations on the rights of children. The young plaintiffs in the 2015 *Juliana* case, then between the ages of 8 and 19, charged that the US government's inaction on addressing climate change violated their constitutional right to life, liberty, and property. The case is still being played out. Whether or not the legal outcome is successful for the young plaintiffs, *Juliana* has drawn considerable public attention to the issue of climate change and the rights of children to government protection from climate change.[45]

There is a reciprocal relationship between social initiatives that ensure rights and foster equality and the climate solutions outlined in the four phases above. The social initiatives are not only effective in accelerating solutions, but they themselves also reduce emissions and the impacts of climate change, as in the case of Indigenous peoples' forest tenure, education of women and young girls, and child-led interventions. Conversely, climate solutions benefit society in many ways: clean, renewable energy brings clean air, better health, and affordable energy, while improved agricultural practices and protected ecosystems lead to a more resilient food system, more nutritious foods, good jobs, storm protection, clean water, and other improvements to society.

Climate solutions generate large economic returns. Even without accounting for the many trillions of healthcare savings from lower air pollution and avoided climate damages, according to the Drawdown Review, the more aggressive scenario for keeping global temperatures below the

1.5°C target could result in an estimated global net economic saving of $145 trillion over the lifetimes of these solutions. In addition, the health co-benefits of curbing fossil fuel emissions are sizeable, even when considering only one air pollutant and a limited number of health outcomes. The World Bank estimated the global cost associated with health damage from ambient air pollution (considering only particulate matter [PM$_{2.5}$] and a limited number of health effects) to be $5.7 trillion in 2016. In a scenario where fossil fuel emissions are reduced to zero, these costs would be realized as savings.[46]

As an example of cost-effectiveness, solar panels and wind turbines have much lower operation and maintenance costs than the fossil fuel power plants they replace, so their estimated net lifetime costs come in far below the net initial cost of implementing them. As a result, globally, onshore wind turbines and utility-scale solar panels would save an estimated $8.5 trillion and $25 trillion over their lifetimes, respectively. Some of the other most cost-effective solutions include distributed solar installations ($13 trillion), improving building insulation ($23 trillion), and electric cars ($16 trillion). Eliminating food waste would save another $1 trillion per year.[44]

These numbers tell us that doing the right thing for our children's health and future is also the right thing for the global economy.

INDIVIDUAL ACTIONS

While the path forward is physically and economically realistic, the biggest challenge is bringing political possibility into alignment with these other realities. We each have a role to play here. Our task is easier knowing that the means are at hand—available now—and being deployed in various settings around the world. We are also buoyed by the awareness that meaningful health and economic benefits will be gained for children and adults alike, starting immediately and growing over time. And we are motivated by the urgency of the situation faced by our children. These factors incentivize us to make needed personal sacrifices of a material or economic nature. It is encouraging that the vast majority of people in a recent survey across 17 countries in North America, Europe and Asia—including about three-quarters of Canadians and Americans—say they would be willing to make at least some changes to how they live and work to help tackle the problem. It is worth noting, however, that some of the individual actions discussed in this chapter (such as purchasing an electric or hybrid car or choosing to live in a walkable city) may not be feasible for low-income populations.[47]

The value of individual action has been the subject of much debate. The concern has been that a focus on voluntary individual action may take the pressure off governments to hold corporate polluters accountable. That it might distract us from engaging in collective political action, which is far more effective in altering our current trajectory. In the magazine of the Sierra Club, Editor Jason Mark writes that taking personal responsibility for climate change and collective political action are both essential: "Living in accord with one's political vision is a way of laying the foundations for the world we want to see—to engage in a kind of 'prefigurative politics' that makes the future into the now. . . . Individual lifestyle changes can act as a kind of alloy that strengthens political activism. To do the difficult work of walking more lightly on the planet is to bind commitment to conviction."[48]

Individual carbon footprints vary dramatically according to personal lifestyles, behaviors, and actions. Following the order of solutions above, we see that there is much each of us can do in our own lives to reduce emissions. When we question whether our choices can really make a difference, as Katharine Hayhoe, atmospheric scientist and professor of political science, puts it, "Climate change isn't a giant boulder we're trying to roll uphill all by ourselves: there are millions of hands on it."[49]

Homes are a major source of GHG emissions. Roughly 20% of US energy-related GHG emissions stem from heating, cooling, and powering households. As homeowners we can reduce our domestic footprint by choose a utility company that generates its power primarily from wind or solar and installing solar panels where possible. Many utilities offer free energy audits and recommendations for improvements. Sealing drafts and adequately insulating our homes makes them more energy efficient: in some cases, the improvements are eligible for tax credits. If renting, we can lobby the landlord to take these steps and save money at the same time.[50]

Most of us waste considerable amounts of electricity in our homes. We can replace conventional incandescent lightbulbs with LEDs that use up to 80% less energy and are also cheaper in the long run. Other remedies include replacing inefficient appliances with new efficient ones; unplugging or powering down computers, devices, and appliances when charged (they use energy even when not charging); adjusting thermostats down just 1 degree during the heating season and up 1 degree in summer; not wasting water, which requires electricity to pump; and turning off lights when not in use.

We can eat less meat and dairy and more plant-based foods. Studies indicate that a high-fiber, plant-based diet is better for our children's health and our own. By reducing our consumption of animal protein by half, we can shrink our diet's carbon footprint by more than 40%. Going vegetarian is the most effective action and is on the increase. Other remedies are to eat food that is grown locally and in season and to cut down on food waste by only buying the food we can eat. Composting food scraps turns biodegradable food into soil and eliminates waste.[51]

We have many choices in the way we consume manufactured products. Through our purchasing power we can reduce our carbon footprint and also help shift the market to climate friendly products. For example, given that hydrofluorocarbons used as refrigerants are highly potent GHGs, when buying a new fridge or air conditioner, we can choose those appliances that use natural refrigerants, such as propane and ammonium; we can also dispose of our old appliances by recycling. By avoiding single-use plastic cutlery, shopping bags, and drinking bottles, instead using our own reusable shopping bags and water bottles, we save energy and reduce pollution. Reusing and recycling plastics and buying plant-based bioplastics are further steps toward building our own circular economy. By not buying clothing that quickly gets thrown away, we save energy and reduce waste.

Choosing to buy or lease a hybrid or electric car is a smart move for people who have the resources. For some it may also be possible to choose to live in a walkable, smart-growth city or town with bike lanes and high-quality public transportation. But all of us, regardless of our economic resources, can advocate for protected bike lanes and improved mass transit in our towns and cities. We will gain health benefits from increased physical activity and cleaner air, and we will spend less money on fuel. We can use video conferencing for meetings, replacing at least one-third of business trips with video meetings. Vacationing in local destinations and using trains instead of planes are ways to cut down on air travel. If flying is necessary, we can choose to purchase an offset credit from activities that do not contribute significantly to social or environmental harms: the funds support projects that reduce emissions of GHGs or increase carbon sequestration.

We can create and protect green spaces in our communities. Green spaces are carbon sinks and are also necessary for children's optimal development and our health. Whether we plant a single tree in the garden or a whole wood, or just add plants to the windowsill or balcony, we are contributing. We can choose not to replace the grass in our own green space with paving or artificial turf. By joining local conservation organizations, we help to protect and conserve green spaces like local parks, ponds, and community gardens. By joining organizations dedicated to conserving and protecting forests, land, coastal wetlands, and oceans and restoring them when degraded, we help strengthen natural carbon sinks and protect valuable ecosystems at the same time.[52]

Research shows that our individual actions can influence others to follow. When people see their friends and neighbors—people like them—taking action to save energy, they are incentivized to do the same. For example, a study in Connecticut found that residents were much more likely to install solar panels if they were already installed in their zip code and particularly on their street. And a survey of more than 400 people showed that, of the respondents who knew someone who had given up flying because of climate change, around half of them said they flew less because of that example. We can further speed these needed cultural shift by telling friends, family, and neighbors about our choices and why they matter to the climate, their health, and the health and future of their children. We can also personally divest from fossil fuel and pressure pension funds, universities, religious institutions, and other entities with which we are associated to divest.[53]

Government engagement is critical. Our individual choices and behaviors, cumulatively, make a difference. But, by themselves, they are not sufficient to make the drastic cut in emissions necessary to limit climate change to 1.5°C: action by government is essential. Here, each of us can play a part as members of the electorate and as active citizens to make our voices heard by those in power. In the words of environmentalist and former US vice president, Al Gore: "Use your voice, use your vote, use your choice."[54]

According to Rachel Licker, Senior Climate Scientist at the Union of Concerned Scientists, "One of the most important things people can do is hold government accountable. From city council members and county commissioners all the way up to the president, let leaders know this issue is important to you and demand they use the latest scientific data to implement a plan that will reduce emissions and strengthen resiliency in communities already being harmed by climate change. Join up with a community group organizing for change that will be impactful and equitable."[55]

We can find out the names of our representatives and ensure they are making good decisions. We have several avenues for engaging with them. One is to educate them about the many health benefits of climate action, especially for children; the climate solutions available now; and the relevance of that information to their decision-making. And that the solutions are not just about climate: they provide clean air, create jobs, and help vulnerable communities cope with the present impacts of climate change and make them more resilient in the face of new challenges. All of these benefits help adults and children alike. At the same time, Neil Weissman, Dean of Dickinson College in the United States, says we have a stick to wield and should use it. "We must change the political and economic calculus of public officials and business leaders with this message: inaction or insufficient action on climate change is hazardous to your careers."[55]

We can amplify our voices a million-fold by joining and supporting organizations and campaigns that focus on climate action and climate justice. We met some of these in Chapter 5: WE ACT, Fridays for Future, the International Youth Climate Movement, Zero Hour, Extinction Rebellion, School Strike for Climate, Earth Uprising, Mom's Clean Air Force, Natural Resources Defense Council, and the Sierra Club. And there are many more.

Individually and as members of organizations we can reach out to others on behalf of children, especially the most vulnerable. As Aaron Bernstein, a pediatrician at Boston Children's Hospital and Harvard University, says: "[I]f only every parent in America knew that climate action was essential to the well-being of their child and family, we would have no political discourse, no debates about the science, no concerns about the course of action." For "parent," we can read "parents, grandparents, aunts, uncles, and anyone who cares for or cares about children and their future."[56]

It will take a concerted effort on all our parts to move governments and markets to go beyond pronouncements of ambitious goals for addressing climate change to the difficult concrete steps needed to implement them while overcoming partisan politics and other obstacles. I am confident that, armed with the facts and motivated by common concern for children, we can do this. The task may seem Sisyphean, but as Katharine Hayhoe reminded us, we aren't going it alone. Janti Soeripto, President and CEO of Save the Children, adds a central principle: We must stand up for our planet and children's futures.[57]

CONCLUSION

It is often said that we should take the future as seriously as we take the present. We have seen up close the present harm being inflicted on children by fossil fuel pollution and climate change. We have come to understand the dire future that will await them if we don't do everything in our collective and individual power to protect their health and well-being now and in the future. Our response to the crisis must be holistic, aggressive, and immediate, simultaneously cutting emissions of fossil fuel pollution and GHGs, supporting natural carbon sinks, strengthening society, and protecting children from the risks they face today. Fairness should be our guiding principle so that we ensure that those children most at risk and most vulnerable are protected. In that way, all children will receive the benefits of a healthy environment and a sustainable future. We have the tools at hand, both as a society and as individuals, to make a far better world for our children and the generations to come. Let us have the wisdom to act.

NOTES

CHAPTER 1

1. In the interests of readability, notes and sources cited within each paragraph are listed in sequential order at the end of the paragraph. I have not cited sources of information that is readily and publicly available.
2. Klein, Naomi. *This Changes Everything*. New York: Simon & Schuster Paperbacks, 2014.
3. Broecker, Wallace S. "Climatic Change: Are We on the Brink of a Pronounced Global Warming?" Science 189 (August 8 1975): 460–463.
4. Ritchie, Hannah, and Max Roser. "Fossil Fuels." 2017. https://ourworldindata.org/fossil-fuels#global-fossil-fuel-consumption.
5. U.S. Energy Information Administration. "U.S Energy Facts Explained." May 14, 2021. https://www.eia.gov/energyexplained/us-energy-facts/; Comstock, Owen. "Nonfossil Fuel Sources Accounted for 21% of U.S. Energy Consumption in 2020." U.S. Energy Information Administration, 2021. https://www.eia.gov/todayinenergy/detail.php?id=48576; U.S. Energy Information Administration. "U.S. Renewable Energy Consumption Surpasses Coal for the First Time in over 130 Years." May 28, 2020. https://www.eia.gov/todayinenergy/detail.php?id=43895
6. International Energy Agency. "The Future of Petrochemicals." 2018. https://www.iea.org/reports/the-future-of-petrochemicals; Hamilton, Lisa Anne, Steven Feit, Carroll Muffet, Matt Kelso, Samantha Malone Rubright, Courtney Bernhardt, Eric Schaeffer, et al. "Plastic & Climate: The Hidden Costs of a Plastic Planet." Center for International Environmental Law, 2019. https://www.ciel.org/wp-content/uploads/2019/05/Plastic-and-Climate-FINAL-2019.pdf; UNICEF. "The Impacts of Climate Change Put Almost Every Child at Risk." 2021. https://www.unicef.org/stories/impacts-climate-change-put-almost-every-child-risk.
7. The World Counts. "Tons of CO_2 Emitted into the Atmosphere." 2022. https://www.theworldcounts.com/challenges/climate-change/global-warming/global-co2-emissions/story.
8. U.S. Energy Information Administration. "Energy and the Environment Explained." 2020. https://www.eia.gov/energyexplained/energy-and-the-environment/where-greenhouse-gases-come-from.php.
9. Betts, Richard. "Met Office: Atmospheric CO_2 Now Hitting 50% Higher Than Pre-Industrial Levels." Carbon Brief, 2021. https://www.carbonbrief.org/met-office-atmospheric-co2-now-hitting-50-higher-than-pre-industrial-levels;

International Energy Agency. "Global CO2 Emissions in 2019." 2020. https://www.iea.org/articles/global-co2-emissions-in-2019

10. International Energy Agency. "Global CO2 Emissions Rebounded To Their Highest Level in History in 2021." March 8, 2022. https://www.iea.org/news/global-co2-emissions-rebounded-to-their-highest-level-in-history-in-2021
Rivera, Alfredo, Kate Larsen, Hannah Pitt, and Shweta Movalia. "Preliminary US Greenhouse Gas Emissions Estimates for 2021." Rhodium Group, 2022. https://rhg.com/research/preliminary-us-emissions-2021/
International Energy Agency. "Global Energy Review: CO2 Emissions in 2021." March 2022. https://www.iea.org/reports/global-energy-review-co2-emissions-in-2021-2

11. University of Colorado at Boulder. "US Oil and Gas Methane Emissions 60 Percent Higher than Estimated: High Emissions Findings Undercut the Case That Gas Offers Substantial Climate Advantage over Coal." ScienceDaily, 2018. https://www.sciencedaily.com/releases/2018/06/180621141154.htm; International Energy Agency. "Methane from Oil & Gas." 2020. https://www.iea.org/reports/methane-tracker-2020/methane-from-oil-gas; Volcovici, Valerie. "Exclusive U.S., EU Line Up over 20 More Countries for Global Methane Pact." Reuters, October 11, 2021. https://www.reuters.com/business/environment/exclusive-us-eu-line-up-over-20-more-countries-global-methane-pact-2021-10-11/.

12. Ritchie, Hannah, and Max Roser. "CO_2 Emissions by Fuel." 2020. https://ourworldindata.org/emissions-by-fuel; Pandey, Sudhanshu, Ritesh Gautam, Sander Houweling, Hugo Denier van der Gon, Pankaj Sadavarte, Tobias Borsdorff, Otto Hasekamp, et al. "Satellite Observations Reveal Extreme Methane Leakage from a Natural Gas Well Blowout." Proc Nat Acad Sci USA 116, 52 (2019): 26376–26381. https://doi.org/https://doi.org/10.1073/pnas.1908712116; The Climate Reality Project. "3 Big Myths About Natural Gas and Our Climate." 2018. https://www.climaterealityproject.org/blog/3-big-myths-about-natural-gas-and-our-climate.

13. Hamilton et al. "Plastic & Climate: The Hidden Costs."

14. CDP. "New Report Shows Just 100 Companies Are Source of over 70% of Emissions." CDP Worldwide, July 10, 2017. https://www.cdp.net/en/article les/media/new-report-shows-just-100-companies-are-source-of-over-70-of-emissions.

15. Andreoni, Manuela. "Amazon Deforestation Soars to 15-Year High." New York Times, November 19, 2021. https://www.nytimes.com/2021/11/19/world/americas/brazil-amazon-deforestation.html.
World Wildlife Fund. "An Area Roughly the Size of California—166,000 Square Miles—Lost to Deforestation Worldside Between 2004 and 2017." January 13, 2021. https://www.worldwildlife.org/press-releases/an-area-roughly-the-size-of-california-166-000-square-miles-lost-to-deforestation-worldwide-between-2004-and-2017; World Wildlife Fund. "Deforestation and Forest Degradation." 2021. https://www.worldwildlife.org/threats/deforestation-and-forest-degradation; Butler, Rhett A. "Amazon Destruction." December 4, 2020. https://rainforests.mongabay.com/amazon/amazon_destruction.html;

16. Gatehouse, Gabriel. "Deforested Parts of Amazon 'Emitting More CO_2 Than They Absorb'." February 11, 2020. https://www.bbc.com/news/science-environment-51464694.

17. Zaitchik, Alexander. "Rainforest on Fire." The Intercept, 2019. https://theinterc ept.com/2019/07/06/brazil-amazon-rainforest-indigenous-conservation-agrib usiness-ranching/; World Wildlife Fund. "What Animals Live in the Amazon? And 8 Other Amazon Facts." 2017. https://www.worldwildlife.org/stories/what-animals-live-in-the-amazon-and-8-other-amazon-facts.

18. Klimont, Zbigniew, Kaarle Kupiainen, Chris Heyes, Pallav Purohit, Janusz Cofala, Peter Rafaj, Jens Borken-Kleefeld, and Wolfgang Schöpp. "Global Anthropogenic Emissions of Particulate Matter Including Black Carbon." Atmospheric Chemistry Physics 17 (July 17, 2017): 8681–8723. https://doi.org/ 10.5194/acp-17-8681-2017.

19. Shaddick, G., M. L. Thomas, Mudu, G. Ruggeri, and S. Gumy. "Half the World's Population Are Exposed to Increasing Air Pollution." NPJ Climate Atmospheric Sci. 3, 23 (2020). https://doi.org/10.1038/s41612-020-0124-2.

20. Tiseo, Ian. "Annual Pm2.5 Particulate Matter Emissions in the United States from 1990 to 2020." Statista, 2021. https://www.statista.com/statistics/501 298/volume-of-particulate-matter-2-5-emissions-us/; Clay, Karen, and Nicholas Z. Muller. "Recent Increases in Air Pollution: Evidence and Implications for Mortality." National Bureau of Economic Research, 2019. https://www.nber.org/ papers/w26381.

21. IPCC. *Global Warming of 1.5°C: An IPCC Special Report on the Impacts of Global Warming of 1.5°C Above Pre-Industrial Levels and Related Global Greenhouse Gas Emission Pathways, in the Context of Strengthening the Global Response to the Threat of Climate Change, Sustainable Development, and Efforts to Eradicate Poverty.* Geneva: World Meteorological Organization, 2018.
IPCC. "Climate Change 2021: The Physical Science Basis." Intergovernmental Panel on Climate Change. 2022. https://www.ipcc.ch/report/sixth-assessment-report-working-group-i/;

22. Gavin Schmidt Is Quoted in NASA Earth Observatory. "Earth's Global Warming Trend Continues: 2020 Tied for Warmest Year on Record." NASA, 2021. https:// www.nasa.gov/press-release/2020-tied-for-warmest-year-on-record-nasa-analysis-shows; IPCC. "Summary for Policymakers." World Meteorological Organization, 2018. https://www.ipcc.ch/sr15/chapter/spm/.

23. Haines, Andy. "Health Co-Benefits of Climate Action." Lancet Planetary Health 1, 1 (2017): E4–E5. https://doi.org/https://doi.org/10.1016/ S2542-5196(17)30003-7.

24. Sherwood, S. C., M. J. Webb, J. D. Annan, K. C. Armour, M. Forster, J. C. Hargreaves, G. Hegerl, et al. "An Assessment of Earth's Climate Sensitivity Using Multiple Lines of Evidence." Rev Geophysics 58, 4 (2020). https://doi.org/ 10.1029/2019RG000678; Guest Authors. "Why Low-End 'Climate Sensitivity' Can Now Be Ruled Out." August 7, 2020. https://yaleclimateconnections.org/ 2020/08/why-low-end-climate-sensitivity-can-now-be-ruled-out/.

25. IPCC. *Global Warming of 1.5°C.*; James Hansen is quoted in Watts, Jonathan. "We Have 12 Years to Limit Climate Change Catastrophe, Warns UN." The Guardian, 2018. https://www.theguardian.com/environment/2018/oct/08/global-warm ing-must-not-exceed-15c-warns-landmark-un-report.

26. IPCC. "Summary for Policymakers"; IPCC. *Global Warming of 1.5°C*; IPCC. "Climate Change 2021: The Physical Science Basis Summary for Policymakers." 2021. https://www.ipcc.ch/sr15/chapter/spm/ António Guterres is quoted in United Nations. "Secretary-General Calls Latest IPCC Climate Report 'Code Red

for Humanity', Stressing 'Irrefutable' Evidence of Human Influence." August 9, 2021. https://www.un.org/press/en/2021/sgsm20847.doc.htm.

27. Birol, Fatih. "COP26 Climate Pledges Could Help Limit Global Warming to 1.8°C, but Implementing Them Will Be the Key." International Energy Agency, November 4, 2021. https://www.iea.org/commentaries/cop26-climate-pledges-could-help-limit-global-warming-to-1-8-c-but-implementing-them-will-be-the-key.

28. McKibben, Bill. "Why We Need to Keep 80% of Fossil Fuels in the Ground." 350.org, 2016. https://350.org/why-we-need-to-keep-80-percent-of-fossil-fuels-in-the-ground/.

29. Cohen, Judah, Karl Pfeiffer, and Jennifer A. Francis. "Warm Arctic Episodes Linked with Increased Frequency of Extreme Winter Weather in the United States." *Nature Communications* 9, 869 (2018). https://doi.org/10.1038/s41467-018-02992-9; Lindsey, Rebecca. "Understanding the Arctic Polar Vortex." Climate.gov, 2022. https://www.climate.gov/news-features/understanding-climate/understanding-arctic-polar-vortex.

30. Union of Concerned Scientists. "CO_2 and Ocean Acidification: Causes, Impacts, Solutions." 2019. https://www.ucsusa.org/resources/co2-and-ocean-acidification.

31. Cheng, L., J. Abraham, J. Zhu, et al. "Record-Setting Ocean Warmth Continued in 2019." Adv Atmos Sci 37 (2020): 137–142. https://doi.org/10.1007/s00376-020-9283-7; Lijing Cheng and John Abraham are quoted in Brackett, Ron. "World's Oceans Warming at Rate Equal to Five Hiroshima Bombs Dropping Every Second." The Weather Channel, 2020. https://weather.com/science/environment/news/2020-01-14-world-oceans-warming-five-hiroshima-bombs-a-second; McKibben. "Why We Need."

32. The Ocean Portal Team. "Ocean Acidification." Smithsonian. https://ocean.si.edu/ocean-life/invertebrates/ocean-acidification; NOAA Pacific Marine Environmental Laboratory. "What Is Ocean Acidification?" 2018. https://www.pmel.noaa.gov/co2/story/What+is+Ocean+Acidification%3F.

33. Porteus, Cosima S., Peter C. Hubbard, Tamsyn M. Uren Webster, Ronny van Aerle, Adelino V. M. Canário, Eduarda M. Santos, and Rod W. Wilson. "Near-Future CO_2 Levels Impair the Olfactory System of a Marine Fish." Nature Climate Change 8 (2018): 737–743. https://doi.org/10.1038/s41558-018-0224-8; The Ocean Portal Team. "Ocean Acidification"; NOAA Pacific Services Center. "Coral Reef Risk Outlook." January 2, 2012. https://sos.noaa.gov/catalog/datasets/coral-reef-risk-outlook/.

34. Berardelli, Jeff. "How Climate Change Is Making Hurricanes More Dangerous." Yale Climate Connections, 2019. https://yaleclimateconnections.org/2019/07/how-climate-change-is-making-hurricanes-more-dangerous/; Henson, Bob, and Jeff Masters. "Northeast Pummeled with Colossal Flooding, Destructive Tornadoes." September 2, 2021. https://yaleclimateconnections.org/2021/09/northeast-pummeled-with-colossal-flooding-destructive-tornadoes/.

35. Lindsey, Rebecca. "Climate Change: Global Sea Level." January 25, 2021. https://www.climate.gov/news-features/understanding-climate/climate-change-global-sea-level; Krajick, Kevin. "Fossil Plants at Bottom of the Greenland Ice Sheet Warn of Future Melting." Lamont-Doherty Earth Observatory, 2021. https://lamont.columbia.edu/news/fossil-plants-bottom-greenland-ice-sheet-warn-future-melting;

Joughin, Ian, Benjamin E. Smith, and Brooke Medley. "Marine Ice Sheet Collapse Potentially Under Way for the Thwaites Glacier Basin, West Antarctica." Science 344, 6185 (2014): 735–738. https://doi.org/10.1126/science.1249 055; C40 Cities. "Staying Afloat: The Urban Response to Sea Level Rise." 2018. https://www.c40.org/other/the-future-we-don-t-want-staying-afloat-the-urban-response-to-sea-level-rise; University of Colorado at Boulder. "Thwaites Science Discussed at AGU Press Conference." International Thwaites Glacier Collaboration, 2021. https://thwaitesglacier.org/news/thwaites-science-discus sed-agu-press-conference.

36. The Copernicus Programme. "Wildfires Wreaked Havoc in 2021, CAMS Tracked Their Impact." 2021. https://atmosphere.copernicus.eu/wildfires-wreaked-havoc-2021-cams-tracked-their-impact; Gibbens, Sarah. "Wildfire Smoke Blowing Across the U.S. Is More Toxic Than We Thought." National Geographic, 2021. https://www.nationalgeographic.com/environment/article/wildfire-smoke-blowing-across-country-more-toxic-than-we-thought; Szekely, Peter, and Steve Gorman. "Western Wildfire Smoke Causes Cross-Country Air Pollution." Reuters, July 21, 2021. https://www.reuters.com/world/us/smoke-us-west-wildfi res-leaves-easterners-gasping-2021-07-20/.

37. United Nations Environment Programme. "Spreading Like Wildfire: The Rising Threat of Extraordinary Landscape Fires." 2022. https://www.unep.org/resour ces/report/spreading-wildfire-rising-threat-extraordinary-landscape-fires.

38. Food and Agriculture Organization. "Land & Water." 2021. http://www.fao.org/ land-water/water/drought/droughtandag/en/; Williams, A. Park, Edward R. Cook, Jason E. Smerdon, Benjamin I. Cook, John T. Abatzoglou, Kasey Bolles, Seung H. Baek, Andrew M. Badger, and Ben Livneh. "Large Contribution from Anthropogenic Warming to an Emerging North American Megadrought." Science 368, 6488 (2020): 314–318. https://doi.org/10.1126/science.aaz9600.

39. NOAA National Centers for Environmental Information. "Billion-Dollar Weather and Climate Disasters: Overview." 2021. https://www.ncdc.noaa.gov/billions/ overview;

40. IPCC. "Climate Change: A Threat to Human Wellbeing and Health of the Planet. Taking Action Now Can Secure Our Future." February 28, 2022. "https://www. ipcc.ch/2022/02/28/pr-wgii-ar6.

41. Bergeron, Louis. "Discovering Mammals Cause for Worry." Stanford News, 2009. https://news.stanford.edu/news/2009/february11/numa-021109.html; Thomas, Chris D., Alison Cameron, Rhys E. Green, Michel Bakkenes, Linda J. Beaumont, Yvonne C. Collingham, Barend F. N. Erasmus, et al. "Extinction Risk from Climate Change." Nature 427, 6970 (2004/01/01 2004): 145–148. https://doi. org/10.1038/nature02121.

42. IPBES. "Summary for Policymakers of the Global Assessment Report on Biodiversity and Ecosystem Services." 2019. https://www.ipbes.net/global-ass essment.

43. World Wildlife Fund. "Living Planet Report 2020: Bending the Curve of Biodiversity Loss." 2020. https://www.zsl.org/sites/default/files/LPR%202 020%20Full%20report.pdf.

44. Kolbert, Elizabeth. *The Sixth Extinction: An Unnatural History*. New York: Henry Holt, 2014; Dreifus, Claudia. "Chasing the Biggest Story on Earth." New York Times, 2014. https://www.nytimes.com/2014/02/11/science/the-sixth-extinct ion-looks-at-human-impact-on-the-environment.html.

45. Pearce, Fred. "As Climate Change Worsens, a Cascade of Tipping Points Looms." December 5, 2019. https://e360.yale.edu/features/as-climate-changes-wors ens-a-cascade-of-tipping-points-looms; James Hansen is quoted in McSweeney, Robert. "Nine 'Tipping Points' That Could Be Triggered by Climate Change." CarbonBrief, 2020. https://www.carbonbrief.org/explainer-nine-tipping-points- that-could-be-triggered-by-climate-change.

46. Scott, Michon, and Rebecca Lindsey. "2019 Arctic Sea Ice Extent Ties for Second- Lowest Summer Minimum on Record." Climate.gov, 2019. https://www.climate. gov/news-features/featured-images/2019-arctic-sea-ice-extent-ties-second-low est-summer-minimum-record; Hugonnet, Romain, Robert McNabb, Etienne Berthier, Brian Menounos, Christopher Nuth, Luc Girod, Daniel Farinotti, et al. "Accelerated Global Glacier Mass Loss in the Early Twenty-First Century." Nature 592 (2021): 726–731. https://doi.org/10.1038/s41586-021-03436-z.

47. climatetippingpoints.info. "What Are Climate Tipping Points?" 2019. https:// climatetippingpoints.info/what-are-climate-tipping-points/.

48. Turetsky, Merritt R., Benjamin W. Abbott, Miriam C. Jones, Katey Walter Anthony, David Olefeldt, Edward A. G. Schuur, Charles Koven, et al. "Permafrost Collapse Is Accelerating Carbon Release." Nature 569 (2019): 32–34. https:// doi.org/https://doi.org/10.1038/d41586-019-01313-4; Shea, Shannon Brescher. "Defrosting the World's Freezer: Thawing Permafrost." The Office of Science, 2017. https://www.energy.gov/science/articles/defrosting-world-s-freezer-thaw ing-permafrost; Schuur, T. "Permafrost and the Global Carbon Cycle." NOAA Arctic Program, 2019. https://arctic.noaa.gov/Report-Card/Report-Card-2019/ ArtMID/7916/ArticleID/844/Permafrost-and-the-Global-Carbon-Cycle.

49. Shukman, David. "Brazil's Amazon: Deforestation 'Surges to 12-Year High'." BBC, 2020. https://www.bbc.com/news/world-latin-america-55130304; The Climate Reality Project. "How Feedback Loops Are Making the Climate Crisis Worse." 2020. https://www.climaterealityproject.org/blog/how-feedback-loops- are-making-the-climate-crisis-worse; Carlos Nobre is quoted in Kormann, Carolyn. "Deforestation, Agriculture, and Diet Are Fuelling the Climate Crisis." The New Yorker, 2019. https://www.newyorker.com/news/news-desk/deforestation- agriculture-and-diet-are-fuelling-the-climate-crisis.

50. The World Counts. "Tons of CO2 Emitted."

51. McSweeney, Robert. "Nine 'Tipping Points'."

52. Broecker, Wallace S. "Thermohaline Circulation, the Achilles Heel of Our Climate System: Will Man-Made CO_2 Upset the Current Balance?" Science 278, 5343 (November 28, 1997): 1582–1588. https://doi.org/10.1126/scie nce.278.5343.1582; Caesar, L., G. D. McCarthy, D. J. R. Thornalley, N. Cahill, and S. Rahmstorf. "Current Atlantic Meridional Overturning Circulation Weakest in Last Millennium." Nature Geoscience 14 (2021): 118–120; Stefan Rahmstorf is quoted in PIK Press Office. "Gulf Stream System at Its Weakest in over a Millennium." Potsdam Institute for Climate Impact Research, 2021. https:// www.sciencedaily.com/releases/2021/02/210225113357.htm.

53. PIK Press Office. "Gulf Stream System"; University of Exeter. "Atlantic Circulation Collapse Could Cut British Crop Farming." 2020. https://www.exeter. ac.uk/news/featurednews/title_772755_en.html.

54. Stevens, William K. "Scientist at Work: Wallace S. Broecker; Iconoclastic Guru of the Climate Debate." New York Times, 1998. https://www.nytimes.com/1998/ 03/17/science/scientist-at-work-wallace-s-broecker-iconoclastic-guru-of-the- climate-debate.html.

CHAPTER 2

1. The report of the Committee on Pesticides in 1996, chaired by Philip Landrigan, was instrumental in highlighting the vulnerability of the developing infant and child. See National Research Council (US) Committee on Pesticides in the Diets of Infants and Children. *Pesticides in the Diets of Infants and Children.* Washington, DC: National Academies Press, 1993. https://www.ncbi.nlm.nih.gov/books/NBK236275/ doi:10.17226/2126. In congressional testimony during the hearing on the Food Quality Protection Act of 1995, Dr. Landrigan cited the "fundamental tenet of pediatric medicine that children are not just little adults."

2. Zhang, Ying, Peng Bi, and Janet E. Hiller, "Climate Change and Disability-Adjusted Life Years." J Environ Health 70, 3 (2007): 32–36.

3. Valuable sources for the section on the brain include Ackerman, Phillip L. "Predicting Individual Differences in Complex Skill Acquisition: Dynamics of Ability Determinants." J Applied Psychol 77, 5 (1992): 598–614 https://doi.org/10.1037/0021-9010.77.5.598; Grandjean, *Phillippe. "Only One Chance: How Environmental Pollution Impairs Brain Development—and How to Protect the Brains of the Next Generation.*, Oxford University Press, 2013; Stiles, J., and Terry L. Jernigan. "The Basics of Brain Development." Neuropsychol Rev 20, 4 (December 2010): 327–348. doi:10.1007/s11065-010-9148-4; Konkel, Lindsey. "The Brain Before Birth: Using fMRI to Explore the Secrets of Fetal Neurodevelopment." Environ Health Perspect 126, 11 (2018): 112001. doi:10.1289/EHP2268; Zero to Three Annual Report 2020. https://www.zerotothree.org/annualreport2020.

4. Konkel, "The Brain Before Birth."

5. Smyser, C. D., et al. "Longitudinal Analysis of Neural Network Development in Preterm Infants." Cerebral Cortex 20, 12 (2010): 2852–2862;

6. Konkel, "The Brain Before Birth."

7. Stiles, J., and T. L. Jernigan. "The Basics of Brain Development." Neuropsychol Rev 20, 4 (2010): 327–348

8. Mizuno, Y., et al., "Structural Brain Abnormalities in Children and Adolescents with Comorbid Autism Spectrum Disorder and Attention-Deficit/Hyperactivity Disorder." Translat Psychiatry 9, 1 (2019): 332.

9. Rothstein, Peter. "Human Development: Chapter 12: Lung Development." 2004. http://www.columbia.edu/itc/hs/medical/humandev/2004/Chpt12-LungDev.pdf

10. Grandjean, "Only One Chance".

11. Ibid.

12. Perera, Frederica., et al., "DNA Damage from Polycyclic Aromatic Hydrocarbons Measured by Benzo[a]pyrene-DNA Adducts in Mothers and Newborns from Northern Manhattan, the World Trade Center Area, Poland, and China." Cancer Epidemiol Biomarkers Prevent 14, 3 (2005): 709–714.
    Woodruff, T. J., A. R. Zota, J. M. Schwartz. "Environmental chemicals in pregnant women in the United States: NHANES 2003–2004." Environ Health Perspect 119(6) (2011): 878–885. doi:10.1289/ehp.1002727

13. Schiavone, Stefania, Vincent Jacquet, Luigia Trabace, and Karl-Heinz Krause. "Severe Life Stress and Oxidative Stress in the Brain: From Animal Models to Human Pathology. Antioxidant Redox Signal 18, 12 (2013): 1475–90.

14. Grandjean, "Only One Chance."

15. Perera, Frederica, et al., "Biomarkers in Maternal and Newborn Blood Indicate Heightened Fetal Susceptibility to Procarcinogenic DNA Damage." Environ Health Perspect 112, 10 (2004): 1133–1136.

16. Simon, A. Katharina, Georg A. Hollandar, and Andrew McMichael. "Evolution of the Immune System in Humans from Infancy to Old Age." Proc Biol Sci 282, 1821 (2015): 20143085.
17. Nelson, B. R., et al., "Immune Evasion Strategies Used by Zika Virus to Infect the Fetal Eye and Brain." Viral Immunol 33, 1 (2020): 22–37.
18. Smith, Caroline J. "Pediatric Thermoregulation: Considerations in the Face of Global Climate Change." Nutrients 11, 9 (2019): 2010. doi:10.3390/nu11092010.
19. Physicians Committee for Responsible Medicine. "Nutritional Requirements throughout the Life Cycle." December 9, 2020. https://nutritionguide.pcrm.org/nutritionguide/view/Nutrition_Guide_for_Clinicians/1342043/all/Nutritional_Requirements_throughout_the_Life_Cycle
20. Faizan, U., A. S. Rouster. "Nutrition and Hydration Requirements In Children and Adults." [Updated 2021 Sep 2]. In: StatPearls [Internet]. Treasure Island (FL): StatPearls Publishing; 2022 Jan-. Available from: https://www.ncbi.nlm.nih.gov/books/NBK562207/
21. World Health Organization, "Children Are Not Little Adults."
22. Ibid.
23. Perera, F., and J. Herbstman. "Prenatal Environmental Exposures, Epigenetics, and Disease." Reprod Toxicol 31, 3 (2011): 363–373.
24. Olvera Alvarez, H. A., et al., "Early Life Stress, Air Pollution, Inflammation, and Disease: An Integrative Review and Immunologic Model of Social-Environmental Adversity and Lifespan Health." Neurosci Biobehav Rev 92 (2018): 226–242; EPA NIEHS/EPA. "Children's Environmental Health and Disease Prevention Research Centers Impact Report: Protecting Children's Health Where They Live, Learn, and Play," 2017. https://www.epa.gov/sites/production/files/2017-10/documents/niehs_epa_childrens_centers_impact_report_2017_0.pdf.
25. This case has been reviewed in detail in Perera, Frederica. "Science as an Early Driver of Policy: Child Labor Reform in the Early Progressive Era, 1870–1900." Am J Public Health 104, 10 (2014):1862–1871 from which much of the following material is drawn.
26. Brace, "Little Laborers of New York City," C.L. Brace, "Little Laborers of New York City," Harpers New Monthly Magazine 47 (1873): 321–332, 326; Graffenried, Clare de. "Child Labor." Publications of the American Economic Association 5, 2 (1890): 216. Stevens, Alzina Parsons. 1894. "Child Slavery in America, Part I." The Arena 117 (1894).
27. Graffenried, "Child Labor.", p. 216
28. Stevens, "Child Slavery in America.", p.123.
29. Graffenried, "Child Labor."

CHAPTER 3
1. This chapter contains material found in prior publications by the author, particularly Perera, F. "Multiple Threats to Child Health from Fossil Fuel Combustion: Impacts of Air Pollution and Climate Change." Environ Health Perspect 125, 2 (2017): 141–148; https://doi.org/10.1289/EHP299 and Perera, F. "Pollution from Fossil-Fuel Combustion Is the Leading Environmental Threat to Global Pediatric Health and Equity: Solutions Exist." Int J Environ Res Public Health 15, 1 (2018): 16. https://doi.org/10.3390/ijerph15010016; Perera, F., K. Nadeau. "Climate Change Fossil Fuel Pollution, and Children's Health." N Engl J Med 86 (2022): 2303–2314, https://www.nejm.org/doi/full/10.1056/NEJMra2117706

2. Watts, G. "The Health Benefits of Tackling Climate Change." November 25, 2009. https://cdn.who.int/media/docs/default-source/climate-change/the-health-ben efi-ts-of-tackling-climate-change170f8b15-d79b-4165-8032-9382861c76a7. pdf?sfvrsn=31763830_1&download=true.

3. The focus of the following section on air pollution is the particles and gases from fossil fuel combustion. Lead and mercury are both toxic metals released by the burning of coal and oil, but their health impacts are not reviewed here since they have long been established—and long understood by the public—to be harmful to developing brains. However, they certainly deserve mention as important co-pollutants emitted by combustion. Mercury, largely emitted by coal-fired power plants, is rained out of the atmosphere into surface waters where it is transformed into methyl mercury, and bioaccumulates in fish and shellfish, which are consumed by humans, including pregnant women. Exposure in utero, measured by cord blood concentrations of mercury, has been associated with impairment of memory, language, and attention in children who were exposed to methylmercury as fetuses. Research on lead over many decades was among the first to highlight the long-term consequences of early exposure to toxicants, documenting decreased IQ and attention problems in childhood and later delinquent activity in adolescents.

4. UNICEF. "Environment and Climate Change." 2020. https://www.unicef.org/envi ronment-and-climate-change.

5. Vohra, Karn, Alina Vodonos, Joel Schwartz, Eloise A. Marais, Melissa Sulprizio, and Loretta J. Mickley. "Global Mortality from Outdoor Fine Particle Pollution Generated by Fossil Fuel Combustion: Results from GEOS-Chem." Environ Res 195 (April 1, 2021): 110754; European Environment Agency. "Air Quality in Europe: 2020 report." 2020. https://www.eea.europa.eu/publications/air-qual ity-in-europe-2020-report.

6. State of Global Air. "Impacts on Newborns." 2020. https://www.stateofglobal air.org/health/newborns; Schraufnagel, Dean E., John R. Balmes, Clayton T. Cowl, Sara De Matteis, Soon-Hee Jung, Kevin Mortimer, Rogelio Perez-Padilla, et al. "Air Pollution and Noncommunicable Diseases: A Review by the Forum of International Respiratory Societies' Environmental Committee, Part 1: The Damaging Effects of Air Pollution." Chest 155, 2 (2019): 409–416. https://doi. org/10.1016/j.chest.2018.10.042; UNICEF. "Clear the Air for Children." October 2016. https://www.unicef.org/media/60106/file; Drew Shindell is quoted in Duke University, "Cutting Carbon Emissions Sooner Could Save 153 Million Lives." 2018, https://www.sciencedaily.com/releases/2018/03/180319145 243.htm.

7. CDC. "National Center for Health Statistics Centers for Disease Control and Prevention." 2018. http://www.cdc.gov/nchs/fastats/births.htm; World Health Organization, The Partnership for Maternal, Newborn & Child Health. "Born Too Soon: The Global Action Report on Preterm Birth." 2012. http://apps.who.int/ iris/bitstream/handle/10665/44864/9789241503433_eng.pdf;jsessionid=FE368 703F5AD8245495998FDFF57DBF8?sequence=1; Ghosh, Rakesh, Kate Causey, Katrin Burkart, Sara Wozniak, Aaron Cohen, and Michael Brauer. "Ambient and Household PM2.5 Pollution and Adverse Perinatal Outcomes: A Meta-Regression and Analysis of Attributable Global Burden for 204 Countries and Territories." PLoS Medicine 18, 11 (September 2021): e1003852. https://doi.org/10.1371/jour nal.pmed.1003852.

8. CDC. "National Center for Health"; Bekkar, Bruce, Susan Pacheco, Rupa Basu, and Nathaniel DeNicola. "Association of Air Pollution and Heat Exposure with Preterm Birth, Low Birth Weight, and Stillbirth in the US: A Systematic Review." JAMA Network Open 3, 6 (2020): e208243-e43. https://doi.org/10.1001/jama networkopen.2020.8243.

9. Wang, Qiong, Bing Li, Tarik Benmarhnia, Shakoor Hajat, Meng Ren, Tao Liu, Luke D. Knibbs, et al. "Independent and Combined Effects of Heatwaves and PM2.5 on Preterm Birth in Guangzhou, China: A Survival Analysis." Environ Health Perspect 128, 1 (2020): 017006. https://doi.org/doi:10.1289/EHP5117.

10. Johnson, Samantha, and Neil Marlow. "Preterm Birth and Childhood Psychiatric Disorders." Pediatr Res 69, 8 (May 1, 2011): 11–18. https://doi.org/10.1203/PDR.0b013e318212faa0.

11. Shea, E., F. Perera, and D. Mills. "Towards a Fuller Assessment of the Economic Benefits of Reducing Air Pollution from Fossil Fuel Combustion: Per-Case Monetary Estimates for Children's Health Outcomes." Environ Res 182 (March 1, 2020): 109019. https://doi.org/https://doi.org/10.1016/j.envres.2019.109019.

12. CDC. "Prevalence and Characteristics of Autism Spectrum Disorder Among Children Aged 8 Years: Autism and Developmental Disabilities Monitoring Network, 11 Sites, United States, 2018." December 3, 2021. https://www.cdc.gov/mmwr/volumes/70/ss/ss7011a1.htm?s_cid=ss7011a1_w.

13. Chiu, Y. H., H. H. Hsu, B. A. Coull, D. C. Bellinger, I. Kloog, J. Schwartz, R. O. Wright, and R. J. Wright. "Prenatal Particulate Air Pollution and Neurodevelopment in Urban Children: Examining Sensitive Windows and Sex-Specific Associations." Environ Int 87 (February 2016): 56–65. https://doi.org/10.1016/j.envint.2015.11.010.

14. Vishnevetsky, J., D. Tang, H. W. Chang, E. L. Roen, Y. Wang, V. Rauh, S. Wang, et al. "Combined Effects of Prenatal Polycyclic Aromatic Hydrocarbons and Material Hardship on Child IQ." Neurotoxicol Teratol 49 (May-June 2015): 74–80. https://doi.org/http://dx.doi.org/10.1016/j.ntt.2015.04.002; Edwards, S. C., W. Jedrychowski, M. Butscher, D. Camann, A. Kieltyka, E. Mroz, E. Flak, et al. "Prenatal Exposure to Airborne Polycyclic Aromatic Hydrocarbons and Children's Intelligence at 5 Years of Age in a Prospective Cohort Study in Poland." Environ Health Perspect 118, 9 (September 2010): 1326–1331. https://doi.org/http://dx.doi.org/10.1289/ehp.0901070; Perera, F., T. Y. Li, C. Lin, and D. Tang. "Effects of Prenatal Polycyclic Aromatic Hydrocarbon Exposure and Environmental Tobacco Smoke on Child IQ in a Chinese Cohort." Environ Res 114 (April 1, 2012): 40–46. https://doi.org/https://doi.org/10.1016/j.envres.2011.12.011.

15. Perera, F., T. Li, Z. Zhou, T. Yuan, Y. Chen, L. Qu, V. Rauh, Y Zhang, and D. Tang. "Benefits of Reducing Prenatal Exposure to Coal-Burning Pollutants to Children's Neurodevelopment in China." Environ Health Perspect 116, 10 (2008): 1396–1400. https://doi.org/10.1289/ehp.11480.

16. Perera, Frederica, Adiba Ashrafi, Patrick L. Kinney, and David Mills. "Towards a Fuller Assessment of Benefits to Children's Health of Reducing Air Pollution and Mitigating Climate Change Due to Fossil Fuel Combustion." Environ. Res. 172 (May 2019): 55–72. https://doi.org/10.1016/j.envres.2018.12.016.

17. Perera, F. P., D. Tang, S. Wang, J. Vishnevetsky, B. Zhang, D. Diaz, D. Camann, and V. Rauh. "Prenatal Polycyclic Aromatic Hydrocarbon (PAH) Exposure and Child Behavior at Age 6-7 Years." Environ Health Perspect 120, 6 (June 2012): 921–926. https://doi.org/10.1289/ehp.1104315.

18. Kogan, Michael D., Catherine J. Vladutiu, Laura A. Schieve, Reem M. Ghandour, Stephen J. Blumberg, Benjamin Zablotsky, James M. Perrin, et al. "The Prevalence of Parent-Reported Autism Spectrum Disorder Among US Children." Pediatrics 142, 6 (2018): e20174161. https://doi.org/10.1542/peds.2017-4161; Lin, Cheng-Kuan, Yuan-Ting Chang, Fu-Shiuan Lee, Szu-Ta Ghen, and David Christiani. "Association Between Exposure to Ambient Particulate Matters and Risks of Autism Spectrum Disorder in Children: A Systematic Review and Exposure-Response Meta-Analysis." Environ Res Letters 16, 6 (May 2021): 063003. https://doi.org/10.1088/1748 0326/abfcf7;
    Volk, H. E., F. Lurmann, B. Penfold, I. Hertz-Picciotto, R. McConnell. Traffic-related air pollution, particulate matter, and autism. JAMA Psychiatry 70 (1) (2013): 71–77. doi:10.1001/jamapsychiatry.2013.266. PMID: 23404082; PMCID: PMC4019010.
19. Margolis, A. E., J. B. Herbstman, K. S. Davis, V. K. Thomas, D. Tang, Y. Wang, S. Wang, et al. "Longitudinal Effects of Prenatal Exposure to Air Pollutants on Self-Regulatory Capacities and Social Competence." J Child Psychol Psychiatry 57, 7 (July 2016): 851–860. https://doi.org/10.1111/jcpp.12548.
20. Bruha, Lauren, Valentini Spyridou, Georgia Forth, and Dennis Ougrin. "Global Child and Adolescent Mental Health: Challenges and Advances." London J Primary Care 10, 4 (2018): 108–109. https://doi.org/10.1080/17571472.2018.1484332; Yolton, Kimberly, Jane C. Khoury, Jeffrey Burkle, Grace LeMasters, Kim Cecil, and Patrick Ryan. "Lifetime Exposure to Traffic-Related Air Pollution and Symptoms of Depression and Anxiety at Age 12 Years." Environ. Res. 173 (June 2019): 199–206. https://doi.org/10.1016/j.env res.2019.03.005;
    Roberts, Susanna, Louise Arseneault, Benjamin Barratt, Sean Beevers, Andrea Danese, Candice L. Odgers, Terrie E. Moffitt, et al. "Exploration of NO2 and PM2.5 Air Pollution and Mental Health Problems Using High-Resolution Data in London-Based Children from a UK Longitudinal Cohort Study." Psychiatry Res 272 (February 2019): 8–17. https://doi.org/https://doi.org/10.1016/j.psych res.2018.12.050; Perera, F. P., S. Wang, J. Vishnevetsky, B. Zhang, K. J. Cole, D. Tang, V. Rauh, & D. H. Phillips. "Polycyclic aromatic hydrocarbons-aromatic DNA adducts in cord blood and behavior scores in New York city children." Environmental Health Perspectives 119(8) (2011): 1176–1181. https://doi.org/10.1289/ehp.1002705; Perera, F. P., S. Wang, V. Rauh, H. Zhou, L. Stigter, D. Camann, W. Jedrychowski, E. Mroz, R. Majewska. "Prenatal Exposure to Air Pollution, Maternal Psychological Distress, and Child Behavior." Pediatrics 132(5) (2013): e1284–e1294. doi:10.1542/peds.2012-3844. Epub 2013 Oct 7.
21. Peterson, B. S., V. A. Rauh, R. Bansal, X. Hao, Z. Toth, G. Nati, K. Walsh, et al. "Effects of Prenatal Exposure to Air Pollutants (Polycyclic Aromatic Hydrocarbons) on the Development of Brain White Matter, Cognition, and Behavior in Later Childhood." JAMA Psychiatry 72, 6 (June 2015): 531–540. https://doi.org/10.1001/jamapsychiatry.2015.57; Guxens, M., M. J. Lubczynska, R. L. Muetzel, A. Dalmau-Bueno, V. W. V. Jaddoe, G. Hoek, A. van der Lugt, et al. "Air Pollution Exposure During Fetal Life, Brain Morphology, and Cognitive Function in School-Age Children." Biol Psychiatry 84, 4 (August 15, 2018): 295–303. https://doi.org/10.1016/j.biopsych.2018.01.016; Burnor, Elisabeth, Dora Cserbik, Devyn L. Cotter, Clare E Palmer, Hedyeh Ahmadi, Sandrah P Eckel, Kiros Berhane, et al. "Association of Outdoor Ambient Fine Particulate Matter with Intracellular White Matter Microstructural Properties Among Children."

JAMA Netw Open 4, 12 (2021): e2138300. https://doi.org/10.1001/jamanetw orkopen.2021.38300; Peterson, B. S., R. Bansal, S. Sawardekar, C. Nati, E. R. Elgabalawy, L. A. Hoepner, W. Garcia, X. Hao, A. Margolis, F. Perera, V. Rauh. "Prenatal Exposure to Air Pollution is Associated With Altered Brain Structure, Function, and Metabolism in Childhood." J Child Psychol Psychiatry 2022 Feb 14. doi:10.1111/jcpp.13578. Epub ahead of print. PMID: 35165899. "Prenatal Exposure to Air Pollution Is Associated with Altered Brain Structure, Function, and Metabolism in Childhood." J Child Psychol Psychiatry (2022).. https://doi. org/10.1111/jcpp.13578.

22. Volkmar, F. R., and J. M. Wolf. "When Children with Autism Become Adults." World Psychiatry 12, 1 (February 2013): 79–80. https://doi.org/10.1002/ wps.20020; Perera, F., K. Weiland, M. Neidell, and S. Wang. "Prenatal Exposure to Airborne Polycyclic Aromatic Hydrocarbons and IQ: Estimated Benefit of Pollution Reduction." J Public Health Policy 35, 3 (August 2014): 327–336. https://doi.org/10.1057/jphp.2014.14

23. Calderón-Garcidueñas, L., A. Gónzalez-Maciel, R. Reynoso-Robles, R. Delgado-Chávez, P. S. Mukherjee, R. J. Kulesza, R. Torres-Jardón, J. Ávila-Ramírez, R. Villarreal-Ríos. "Hallmarks of Alzheimer disease are Evolving Relentlessly in Metropolitan Mexico City Infants, Children and Young Adults." Environ Res 164 (2018): 475–487. doi:10.1016/j.envres.2018.03.023. Epub 2018 Mar 26. Younan, Diana, Andrew J. Petkus, Keith F Widaman, Xinhui Wang, Ramon Casanova, Mark A. Espeland, Margaret Gatz, et al. "Particulate Matter and Episodic Memory Decline Mediated by Early Neuroanatomic Biomarkers of Alzheimer's Disease." Brain 143, 1 (2019): 289–302. https://doi.org/10.1093/brain/awz348.

24. Ferrante, G., and S. La Grutta. "The Burden of Pediatric Asthma." Front Pediatr 6 (2018): 186. https://doi.org/10.3389/fped.2018.00186; Nicholas, Stephen W., Betina Jean-Louis, Benjamin Ortiz, Mary Northridge, Katherine Shoemaker, Roger Vaughan, Michaela Rome, Geoffrey Canada, and Vincent Hutchinson. "Addressing the Childhood Asthma Crisis in Harlem: The Harlem Children's Zone Asthma Initiative." Am J Public Health 95, 2 (2005): 245–249. https://doi. org/10.2105/AJPH.2004.042705.

25. Harlem Children's Zone. "Harlem Children's Zone: A Look Inside." 2005. https:// hcz.org/wp-content/uploads/2014/04/ALI-Asthma.pdf.

26. Khreis, H., C. Kelly, J. Tate, R. Parslow, K. Lucas, and M. Nieuwenhuijsen. "Exposure to Traffic-Related Air Pollution and Risk of Development of Childhood Asthma: A Systematic Review and Meta-Analysis." Environ Int 100 (March 2017): 1–31. https://doi.org/10.1016/j.envint.2016.11.012; Anenberg, Susan C., Arash Mohegh, Daniel L. Goldberg, Gaige H. Kerr, Michael Brauer, Katrin Burkart, Perry Hystad, et al. "Long-Term Trends in Urban NO2 Concentrations and Associated Paediatric Asthma Incidence: Estimates from Global Datasets." Lancet Planetary Health, 2022. https://doi.org/10.1016/ S2542-5196(21)00255-2.

27. Avol, E. L., W. J. Gauderman, S. M. Tan, S. J. London, and J. M. Peters. "Respiratory Effects of Relocating to Areas of Differing Air Pollution Levels." Am J Respir Crit Care Med 164, 11 (December 1, 2001): 2067–2072. https://doi.org/ 10.1164/ajrccm.164.11.2102005.

28. Horne, B. D., E. A. Joy, M. G. Hofmann, H. Gesteland, J. B. Cannon, J. S. Lefler, D. Blagev, et al. "Short-Term Elevation of Fine Particulate Matter Air Pollution and Acute Lower Respiratory Infection." Am J Respir Crit Care Med 198, 6 (September 15, 2018): 759–766. https://doi.org/10.1164/rccm.201709-1883OC.

Marusic, Kristina. "Kids With Asthma Who Live Near Heavy Air Pollution Face Greater Risk From Coronavirus." Environmental Health News. April 3, 2020. https://www.ehn.org/children-asthma-coronavirus-2645618537.html

29. Tai, A., H. Tran, M. Roberts, N. Clarke, A. M. Gibson, S. Vidmar, J. Wilson, and C. F. Robertson. "Outcomes of Childhood Asthma to the Age of 50 Years." J Allergy Clin Immunol 133, 6 (June 2014): 1572-8.e3. https://doi.org/10.1016/j.jaci.2013.12.1033.

30. CDC. "RSV in Infants and Young Children." 2018. https://www.cdc.gov/rsv/high-risk/infants-young-children.html.

31. Nadeau, K., C. McDonald-Hyman, E. M. Noth, B. Pratt, S. K. Hammond, J. Balmes, and I. Tager. "Ambient Air Pollution Impairs Regulatory T-Cell Function in Asthma." J Allergy Clin Immunol 126, 4 (October 2010): 845–52.e10. https://doi.org/10.1016/j.jaci.2010.08.008.

32. University Hospitals. "Prenatal Exposure to Air Pollution Linked to Sleep Disruption in Toddlers." 2019. https://www.uhhospitals.org/Healthy-at-UH/articles/2019/05/prenatal-exposure-to-air-pollution-linked-to-sleep-disruption-in-toddlers.

33. Helldén, D., C. Andersson, M. Nilsson, K. L. Ebi, P. Friberg, and T. Alfvén. "Climate Change and Child Health: A Scoping Review and an Expanded Conceptual Framework." Lancet Planetary Health 5, 3 (2021): E164–E175. https://doi.org/10.1016/S2542-5196(20)30274-6; Sheffield, E., and J. Landrigan. "Global Climate Change and Children's Health: Threats and Strategies for Prevention." Environ Health Perspect 119, 3 (March 2011): 291–298. https://doi.org/10.1289/ehp.1002233.

34. UNICEF. The Climate Crisis Is a Child Rights Crisis: Introducing the Children's Climate Risk Index. New York: United Nations Childrens Fund, 2021. https://www.unicef.org/media/105376/file/UNICEF-climate-crisis-child-rights-crisis.pdf.

35. Limaye, Vijay. "The Costs of Inaction: The Economic Burden of Fossil Fuels and Climate Change on Health in the U.S." NRDC, May 20, 2021. https://www.nrdc.org/resources/costs-inaction-economic-burden-fossil-fuels-and-climate-change-health-us.

36. Pronczuk, J., and S. Surdu. "Children's Environmental Health in the Twenty-First Century." Ann N Y Acad Sci 1140 (October 2008): 143–154. https://doi.org/10.1196/annals.1454.045; Save the Children. "Legacy of Disasters." 2007. https://resourcecentre.savethechildren.net/node/3986/pdf/3986.pdf; IFRC. "World Disasters Report 2020." 2020. https://reliefweb.int/report/world/world-disasters-report-2020-come-heat-or-high-water-tackling-humanitarian-impacts; Holland, Greg, and Cindy L. Bruyère. "Recent Intense Hurricane Response to Global Climate Change." Climate Dynamics 42, 3 (February 1, 2014): 617–627. https://doi.org/10.1007/s00382-013-1713-0.

37. Torti, Jacqueline. "Floods in Southeast Asia: A Health Priority." J Global Health 2, 2 (2012): 020304-04. https://doi.org/10.7189/jogh.02.020304; UNICEF. "Unless We Act Now: The Impact of Climate Change on Children." 2015. https://www.unicef.org/reports/unless-we-act-now-impact-climate-change-children.

38. Reliefweb. "Many Children Dying and Jobs Being Lost in Asian Floods Says UNISDR." 2011. https://reliefweb.int/report/thailand/many-children-dying-and-jobs-being-lost-asian-floods-says-unisdr; UNICEF USA. "Hurricane Relief Philippines Typhoon Haiyan." 2014. https://www.unicefusa.org/mission/emergencies/hurricanes/2013-philippines-typhoon-haiyan#; UNICEF. "16 Million

Children Affected by Massive Flooding in South Asia, with Millions More at Risk." September 2, 2017. https://www.unicef.org/press-releases/16-million-children-affected-massive-flooding-south-asia-millions-more-risk.

39. Chakrabarti, Rajashri. "The Impact of Superstorm Sandy on New York City School Closures and Attendance." Huffpost, December 24, 2012. https://www.huffpost.com/entry/hurricane-sandy-school-days_b_2360754; Gibbens, Sarah. "Hurricane Sandy, Explained." National Geographic, February 11, 2019. https://www.nationalgeographic.com/environment/article/hurricane-sandy; Pacheco, Susan E. "Hurricane Harvey and Climate Change: The Need for Policy to Protect Children." Pediatr Res 83, 1 (2018): 9–10. https://doi.org/10.1038/pr.2017.280. The New York Times. Flooding From Ida Kills Dozens of People in 4 States. 2021. https://www.nytimes.com/live/2021/09/02/nyregion/nyc-storm

40. United Nations Office for Disaster Risk Reduction. "UNISDR Says the Young Are the Largest Group Affected by Disasters." 2011https://www.undrr.org/news/unisdr-says-young-are-largest-group-affected-disasters; Lai, Betty S., Annette M. La Greca, Ahnalee Brincks, Courtney A. Colgan, Michelle D'Amico, Sarah Lowe, and Mary Lou Kelley. "Trajectories of Posttraumatic Stress in Youths After Natural Disasters." JAMA Network Open 4, 2 (2021): e2036682-e82. https://doi.org/10.1001/jamanetworkopen.2020.36682; Gil-Rivas, Virginia, and Ryan Kilmer. "Children's Adjustment Following Hurricane Katrina: The Role of Primary Caregivers." Am J Orthopsychiatry 83, 2 Pt 3 (April–July 2013): 413–421. https://doi.org/10.1111/ajop.12016.

41. UCLA. "Voices for Katrina's Children." 2006. http://cretscmhd.psych.ucla.edu/nola/volunteer/FoundationReports/voices.pdf.

42. NOAA. "Understanding the Polar Vortex." 2021. https://www.climate.gov/news-features/understanding-climate/understanding-arctic-polar-vortex; Texas Tribune. "At Least 57 People Died in the Texas Winter Storm, Mostly from Hypothermia." 2021. https://www.texastribune.org/2021/03/15/texas-winter-storm-deaths/.

43. NASA. "2020 Tied for Warmest Year on Record, NASA Analysis Shows." 2020. https://www.nasa.gov/press-release/2020-tied-for-warmest-year-on-record-nasa-analysis-shows; Thiery, Wim, Stefan Lange, Joeri Rogelj, Carl-Friedrich Schleussner, Lukas Gudmundsson, Sonia I. Seneviratne, Marina Andrijevic, et al. "Intergenerational Inequities in Exposure to Climate Extremes." Science 374, 6564 (2021): 158–160. https://doi.org/doi:10.1126/science.abi7339.

44. Wang, "Independent and Combined Effects"; Chersich, Matthew F., Minh Duc Pham, Ashtyn Areal, Marjan M. Haghighi, Albert Manyuchi, Callum P. Swift, Bianca Wernecke, et al. "Associations Between High Temperatures in Pregnancy and Risk of Preterm Birth, Low Birth Weight, and Stillbirths: Systematic Review and Meta-Analysis." BMJ 2020, 371 (September2020): m3811. https://doi.org/10.1136/bmj.m3811.

45. Knowlton, K., M. Rotkin-Ellman, G. King, H. G. Margolis, D. Smith, G. Solomon, R. Trent R, and English P. "The 2006 California Heat Wave: Impacts on Hospitalizations and Emergency Department Visits." Environ Health Perspect 117, 1 (January 2009): 61–67. https://doi.org/10.1289/ehp.11594; Bernstein, Aaron S., Shengzhi Sun, Kate R. Weinberger, Keith R. Spangler, Perry E. Sheffield, and Gregory A. Wellenius. "Warm Season and Emergency Department Visits to U.S. Children's Hospitals." Environ Health Perspect 130, 1 (January 2022). https://doi.org/10.1289/EHP8083; Clemens, Vera, Eckart von Hirschhausen, and Jörg M. Fegert. "Report of the Intergovernmental Panel

on Climate Change: Implications for the Mental Health Policy of Children and Adolescents in Europe: A Scoping Review." Eur Child Adolesc Psychiatry (August 2020). https://doi.org/10.1007/s00787-020-01615-3.

46. Isen, Adam, Maya Rossin-Slater, and Reed Walker. "Relationship Between Season of Birth, Temperature Exposure, and Later Life Wellbeing." Proc Nat Acad Sci 114, 51 (2017): 13447–13452. https://doi.org/10.1073/pnas.1702436114.

47. Sheffield, Landrigan, "Global Climate Change."

48. van der Werf, G. R., J. T. Randerson, L. Giglio, T. T. van Leeuwen, Y. Chen, B. M. Rogers, M. Mu, et al. "Global Fire Emissions Estimates During 1997–2016." Earth Syst Sci Data 9, 2 (2017): 697–720. https://doi.org/10.5194/essd-9-697-2017.

49. Congressional Research Service. "Wildfire Statistics." 2020. https://fas.org/sgp/crs/misc/IF10244.pdf; Burke, Marshall, Anne Driscoll, Sam Heft-Neal, Jiani Xue, Jennifer Burney, and Michael Wara. "The Changing Risk and Burden of Wildfire in the United States." Proc Nat Acad Sci 118, 2 (2021): e2011048118. https://doi.org/10.1073/pnas.2011048118; Holm, Stephanie M., Mark D. Miller, and John R. Balmes. "Health Effects of Wildfire Smoke in Children and Public Health Tools: A Narrative Review." J Exposure Sci Environ Epidemiol 31, 1 (January 1, 2021): 1–20. https://doi.org/10.1038/s41370-020-00267-4.

50. Abdo, Mona, Isabella Ward, Katelyn O'Dell, Bonne Ford, Jeffrey R. Pierce, Emily V. Fischer, and James L. Crooks. "Impact of Wildfire Smoke on Adverse Pregnancy Outcomes in Colorado, 2007–2015." Int J Environ Res Public Health 16, 19 (2019): 3720. https://doi.org/10.3390/ijerph16193720; Pratt, J. R., R. W. Gan, B. Ford, S. Brey, J. R. Pierce, E. V. Fischer, and S. Magzamen. "A National Burden Assessment of Estimated Pediatric Asthma Emergency Department Visits That May Be Attributed to Elevated Ozone Levels Associated with the Presence of Smoke." Environ Monit Assess 191, Suppl 2 (June 28, 2019): 269. https://doi.org/10.1007/s10661-019-7420-5.

51. CDC. "Hospitalization Rates and Characteristics of Children Aged <18 Years Hospitalized with Laboratory-Confirmed COVID-19—COVID-NET, 14 States, March 1–July 25, 2020." 2020. https://www.cdc.gov/mmwr/volumes/69/wr/mm6932e3.htm.

52. Holm et al., "Health Effects of Wildfire Smoke."; Yelland, C., Robinson, C. Lock, A. M. La Greca, B. Kokegei, V. Ridgway, and B. Lai. "Bushfire Impact on Youth." J Trauma Stress 23, 2 (April 2010): 274–277. https://doi.org/10.1002/jts.20521.

53. UNICEF, "Unless We Act Now"; World Economic Forum. "How Does Climate Change Make People Poorer?" 2015. https://www.weforum.org/agenda/2015/11/how-does-climate-change-make-people-poorer/.

54. World Health Organization. "Joint Child Malnutrition Estimates." 2020. https://www.who.int/data/gho/data/themes/topics/joint-child-malnutrition-estimates-unicef-who-wb.

55. Bauer, L. "About 14 Million Children in the US Are Not Getting Enough to Eat." Brookings, 2020. https://www.brookings.edu/blog/up-front/2020/07/09/about-14-million-children-in-the-us-are-not-getting-enough-to-eat/.

56. ScienceDaily. "Rising CO_2, Climate Change Projected to Reduce Availability of Nutrients Worldwide." 2019. https://www.sciencedaily.com/releases/2019/07/190718085308.htm.

57. Rocklöv, J., and R. Dubrow. "Climate Change: An Enduring Challenge for Vector-Borne Disease Prevention and Control." Nat Immunol 21, 5 (May 2020): 479–483. https://doi.org/10.1038/s41590-020-0648-y. https://www.ncbi.nlm.nih.

gov/pubmed/32313242; Nelson, B. R., J. A. Roby, W. B. Dobyns, L. Rajagopal, M. Gale, Jr., and K. M. Adams Waldorf. "Immune Evasion Strategies Used by Zika Virus to Infect the Fetal Eye and Brain." Viral Immunol 33, 1 (January/February 2020): 22–37. https://doi.org/10.1089/vim.2019.0082; Siraj, A. S., M. Santos-Vega, M. J. Bouma, D. Yadeta, D. Ruiz Carrascal, and M. Pascual. "Altitudinal Changes in Malaria Incidence in Highlands of Ethiopia and Colombia." Science 343, 6175 (2014): 1154–1158. https://doi.org/10.1126/science.1244325; Acharaya, Bipin K., Chunxiang Cao, Min Xu, Laxman Khanal, Shahid Naeem, and Shreejana Pandit. "Present and Future of Dengue Fever in Nepal: Mapping Climatic Suitability by Ecological Niche Model." Int J Environ Res Public Health 15, 2 (2018): 187. https://doi.org/10.3390/ijerph15020187; CDC. "Lyme Disease Data Tables: Historical Data Reported: Cases of Lyme Disease by State or Locality, 2009–2018." 2016. https://www.cdc.gov/lyme/stats/tables.html.

58. WHO. "The World Malaria Report 2019" at a Glance." 2019. https://www.who. int/news-room/feature-stories/detail/world-malaria-report-2019.

59. CDC. "Lyme Disease." 2019. https://www.cdc.gov/lyme/why-is-cdc-concerned-about-lyme-disease.html.

60. CDC. "Health Concerns Associated with Mold in Water-Damaged Homes After Hurricanes Katrina and Rita—New Orleans Area, Louisiana, October 2005." 2006. https://www.cdc.gov/mmwr/preview/mmwrhtml/mm5502a6.htm.

61. Scidev. "Climate Now Biggest Driver of Migration, Study Finds." 2019. https:// www.scidev.net/global/news/climate-now-biggest-driver-of-migration-study-finds/; European Economic and Social Committee. "Climate Refugees Account for More Than a Half of All Migrants but Enjoy Little Protection." 2020. https:// www.eesc.europa.eu/en/news-media/news/climate-refugees-account-more-half-all-migrants-enjoy-little-protection;
UNICEF. "Climate Mobility and Children." 2019. https://www.unicef.org/global insight/climate-mobility-and-children

62. Yale Climate Connections. "As Wildfires, Flooding, and Hurricanes Grow More Frequent, Climate Migration Begins." September 22, 2020. https://yaleclimate connections.org/2020/09/as-wildfires-flooding-and-hurricanes-grow-more-frequent-climate-migration-begins/; New York Times. "Climate Crisis Migration America." 2020. https://www.nytimes.com/interactive/2020/09/15/magazine/climate-crisis-migration-america.html.

63. Kelley, Colin P., Shahrzad Mohtadi, Mark A. Cane, Richard Seager, and Yochanan Kushnir. "Climate Change in the Fertile Crescent and Implications of the Recent Syrian Drought." Proc Nat Acad Sci 112 (11) (2015): 3241–3246. https://doi.org/ 10.1073/pnas.1421533112; UNHCR. "Syria Emergency." 2021. https://www. unhcr.org/en-us/syria-emergency.html.

64. McLaughlin, K. A., J. Greif Green, M. J. Gruber, N. A. Sampson, A. M. Zaslavsky, and R. C. Kessler. "Childhood Adversities and First Onset of Psychiatric Disorders in a National Sample of US Adolescents." Arch Gen Psychiatry 69, 11 (November 2012): 1151–1160. https://doi.org/10.1001/archgenpsychia try.2011.2277.

65. Zero Hour. "Vote 4 Our Future." 2021. http://thisiszerohour.org; Borenstein, Seth. "'A Planet Full of Ifs'; Jamie Margolin, Kaylah Brathwaite, Max Prestigiacomo, and Komal Kumar Are Quoted in Young People Express Climate Angst." AP News, September 25, 2019. https://apnews.com/article/climate-uni ted-nations-health-us-news-climate-change-2d7ccb5d1b3242b9923deb2ca 197a56b.

66. Harrabin, R. "Climate Change: Young People Very Worried—Survey." BBC News, September 2021. https://www.bbc.com/news/world-58549373; Hickman, Caroline, Elizabeth Marks, Panu Pikhala, Susan Clayton, R. Eric Lewandowski, Elouise E. Mayall, Britt Wray, Catriona Mellor, and Lise van Susteren. "Climate Anxiety in Children and Young People and Their Beliefs About Government Responses to Climate Change: A Global Survey." Lancet Planetary Health 5, 12 (2021): E863–E73. https://doi.org/10.1016/S2542-5196(21)00278-3.
67. New York City Department of Health and Mental Hygiene. "Air Pollution and the Health of New Yorkers: The Impact of Fine Particles and Ozone." April 2011 https://www1.nyc.gov/assets/doh/downloads/pdf/eode/eode-air-quality-impact.pdf
68. Masterson, Victoria. "Climate Change: Half the World's Children Are in High-Risk Areas." World Economic Forum. September 1, 2021. https://www.weforum.org/agenda/2021/09/unicef-children-climate-risk- https://www.weforum.org/agenda/2021/09/unicef-children-climate-risk-index/; Hickman et al., "Climate Anxiety in Children"; Wang et al., "Independent and Combined Effects."
69. McEwen, Bruce S., and Pamela Tucker. "Critical Biological Pathways for Chronic Psychosocial Stress and Research Opportunities to Advance the Consideration of Stress in Chemical Risk Assessment." Am J Public Health 101, S1 (2011): S131–S139. https://doi.org/10.2105/AJPH.2011.300270.
70. Landrigan, J., R. Fuller, N. J. R. Acosta, O. Adeyi, R. Arnold, N. N. Basu, A. B. Baldé, et al. "The Lancet Commission on Pollution and Health." Lancet 391, 10119 (February 3, 2018): 462–512. https://doi.org/10.1016/s0140-6736(17)32345-0.
71. Factor-Litvak, Pam, Beverly Insel, Antonia M. Calafat, Xinhua Liu, Frederica Perera, Virginia A. Rauh, and Robin M. Whyatt. "Persistent Associations Between Maternal Prenatal Exposure to Phthalates on Child IQ at Age 7 Years." PLOS ONE 9, 12 (2014): e114003. https://doi.org/10.1371/journal.pone.0114003; Swan, Shanna H., Katharina M. Main, Fan Liu, Sara L. Stewart, Robin L. Kruse, Antonia M. Calafat, Catherine S. Mao, et al. "Decrease in Anogenital Distance Among Male Infants with Prenatal Phthalate Exposure." Environ Health Perspect 113, 8 (2005): 1056–1061. https://doi.org/10.1289/ehp.8100.
72. Rauh, Virginia, Wanda Garcia, Robin Whyatt, Megan Horton, Dana Barr, and Elan Louis. "Prenatal Exposure to the Organophosphate Pesticide Chlorpyrifos and Childhood Tremor." Neurotoxicology 51 (September 21, 2015). https://doi.org/10.1016/j.neuro.2015.09.004; Rauh, Virginia A., Frederica Perera, Megan K. Horton, Robin M. Whyatt, Ravi Bansal, Xuejun Hao, Jun Liu, et al. "Brain Anomalies in Children Exposed Prenatally to a Common Organophosphate Pesticide." Proc Nat Acad Sci USA 109, 20 (2012): 7871–7876. https://doi.org/10.1073/pnas.1203396109.
73. Yang, S. "Exposure to Flame Retardants Linked to Changes in Thyroid Hormones." 2010. https://news.berkeley.edu/2010/06/21/pbde/; Herbstman, J. B., A. Sjödin, M. Kurzon, S. A. Lederman, R. S. Jones, V. Rauh, L. L. Needham, et al. "Prenatal Exposure to PBDEs and Neurodevelopment." Environ Health Perspect 118, 5 (May 2010): 712–719. https://doi.org/10.1289/ehp.0901340.
74. Perera, F., E. L. R. Nolte, Y. Wang, et al. "Bisphenol A Exposure and Symptoms of Anxiety and Depression Among Inner City Children at 10–12 Years of Age." Environ Res 151 (2016): 195–202. doi:10.1016/j.envres.2016.07.028. https://www.ncbi.nlm.nih.gov/pmc/articles/PMC5071142/.

75. People Over Petro. "News: Will a Push for Plastics Turn Appalachia into the Next Cancer Alley?" The Guardian. 2019. https://peopleoverpetro.org/2019/ 10/17/news-will-a-push-for-plastics-turn-appalachia-into-the-next-cancer-alley/; National Geographic. "A Whopping 91% of Plastic Isn't Recycled." 2018. https://www.nationalgeographic.com/science/article/plastic-produced-recycl ing-waste-ocean-trash-debris-environment; Ellen MacArthur Foundation. "The New Plastics Economy: Rethinking the Future of Plastics." 2016. https://ellen macarthurfoundation.org/the-new-plastics-economy-rethinking-the-future-of-plastics;
Kam Sripada, Aneta Wierzbicka, Khaled Abass, Joan O. Grimalt, Andreas Erbe, Halina B. Röllin, Pál Weihe, Gabriela, Jiménez Díaz, Randolph Reyes Singh, Torkild Visnes, Arja Rautio, Jon Øyvind Odland, and Martin Wagner. "A Children's Health Perspective on Nano- and Microplastics." Environmental Health Perspectives (2022): 130. https://doi.org/10.1289/EHP9086.

76. Perera, F. "Multiple Threats to Child Health from Fossil Fuel Combustion: Impacts of Air Pollution and Climate Change." Environ Health Perspect 125, 2 (February 2017): 141–148. https://doi.org/10.1289/ehp299; Perera, F., and K. Nadeau. "Climate Change, Fossil-Fuel Pollution, and Children's Health." New England Journal of Medicine 386(24) (2022): 2303–2314. doi:10.1056/NEJMra2117706.

CHAPTER 4

1. American Psychological Association. *Publication Manual of the American Psychological Association*, 7th ed. Washington, DC: APA, 2020.

2. Maantay, Juliana. "Mapping Environmental Injustices: Pitfalls and Potential of Geographic Information Systems in Assessing Environmental Health and Equity." Environ Health Perspect 110, 2 (2002): 161–171.

3. UNICEF. "Children Living in Poverty." 2005. https://www.unicef.org/sowc05/ english/poverty.html; Children's Defense Fund. "The State of America's Children 2020: Child Poverty." 2020. https://www.childrensdefense.org/policy/resour ces/soac-2020-child-poverty/; Kaiser Family Foundation. "Poverty Rate by Race/ Ethnicity." 2019. https://www.kff.org/other/state-indicator/poverty-rate-by-raceethnicity/.

4. Dr. Tedros Adhanom Ghebreyesus, Director-General of the WHO, is quoted as saying, "9 out of 10 people worldwide breathe polluted air, but more countries are taking action." World Health Organization, 2018. https://www.who.int/ news/item/02-05-2018-9-out-of-10-people-worldwide-breathe-polluted-air-but-more-countries-are-taking-action;

5. Wu, Jin, Derek Watkins, Josh Williams, Shalini Venugopal Bhagat, Hari Kumar, and Jeffrey Gettleman. "Who Gets to Breathe Clean Air in New Delhi?" New York Times, 2020. https://www.nytimes.com/interactive/2020/12/17/world/asia/ india-pollution-inequality.html.

6. American Lung Association. State of the Air. 2021. https://www.lung.org/ research/sota; Tessum, C. W., J. S. Apte, A. L. Goodkind, N. Z. Muller, K. A. Mullins, D. A. Paolella, S. Polasky, et al. "Inequity in Consumption of Goods and Services Add to Racial-Ethnic Disparities in Air Pollution Exposure." Proc Nat Acad Sci 116(13) (2019): 6001–6006). https://doi.org/10.1073/pnas.1818859 116; Chakraborty, Jayajit, and Paul A Zandbergen. "Children at Risk: Measuring Racial/Ethnic Disparities in Potential Exposure to Air Pollution at School and

Home." J Epidemiol Community Health 61, 12 (2007): 1074–1079. https://doi.org/10.1136/jech.2006.054130.

7. European Environment Agency. "Unequal Exposure and Unequal Impacts: Social Vulnerability to Air Pollution, Noise and Extreme Temperatures in Europe." 2019. https://www.eea.europa.eu/publications/unequal-exposure-and-unequal-impacts

8. Burris, Heather H., Scott A. Lorch, Haresh Kirpalani, DeWayne M. Pursley, Michal A. Elovitz, and Jane E. Clougherty. "Racial Disparities in Preterm Birth in USA: A Biosensor of Physical and Social Environmental Exposures." Arch Dis Child 104, 10 (2019): 931–935. https://doi.org/10.1136/archdischild-2018-316 486; Blencowe, H., J. Krasevec, M. de Onis, R. E. Black, X. An, G. A. Stevens, E. Borghi, et al. "National, Regional, and Worldwide Estimates of Low Birthweight in 2015, with Trends from 2000: A Systematic Analysis." Lancet: Global Health 7, 7 (2019): e849–e860. https://doi.org/10.1016/S2214-109X(18)30565-5.

9. CDC. "Infant Mortality." 2020. https://www.cdc.gov/reproductivehealth/maternalinfanthealth/infantmortality.htm.

10. CDC. "Most Recent National Asthma Data." 2019. https://www.cdc.gov/asthma/most_recent_national_asthma_data.htm; Nicholas, S. W., B. Jean-Louis, B. Ortiz, M. Northridge, K. Shoemaker, R. Vaughan, M. Rome, et al. "Addressing the Childhood Asthma Crisis in Harlem: The Harlem Children's Zone Asthma Initiaitive." Am J Public Health 95, 2 (2005): 245–249. https://doi.org/10.2105/AJPH.2004.042705.

11. Shankardass, K., R. McConnell, M. Jerrett, J. Milam, J. Richardson, and K. Berhane. "Parental Stress Increases the Effect of Traffic-Related Air Pollution on Childhood Asthma Incidence." PNAS 106, 30 (2009): 12406–12411. https://doi.org/10.1073/pnas.0812910106.

12. Bixler, D., A. D. Miller, C. P. Mattison, et al. "SARS-CoV-2-Associated Deaths Among Persons Aged <21 Years—United States, February 12—July 31, 2020." CDC, 2020. https://www.cdc.gov/mmwr/volumes/69/wr/mm6937e4.htm.

13. Zablotsky, B., and J. M. Alford. "Prevalence of Attention-deficit/Hyperactivity Disorder and Learning Disabilities Among U.S. Children Aged 3-17 Years." CDC, 2020. https://www.cdc.gov/nchs/products/databriefs/db358.htm; Abrams, A., M. Goyal, and G. Badolato. "Racial Disparities in Pediatric Mental Health-Related Emergency Department Visits: A Five-Year Multi-Institutional Study." Pediatr Emerg Care (2019) 144(2): 414). https://doi.org/10.1097/PEC.0000000000002 221; Alegria, M., M. Vallas, and A. Pumariega. "Racial and Ethnic Disparities in Pediatric Mental Health." Child Adolesc Psychiatr Clin N Am 19, 4 (2010): 759–774. https://doi.org/10.1016/j.chc.2010.07.001.

14. Centre for Research on the Epidemiology of Disasters. The Human Cost of Natural Disasters. 2015. https://reliefweb.int/sites/reliefweb.int/files/resources/PAND_report.pdf; UNICEF. "2013 Philippines Typhoon Haiyan." UNICEF USA, 2013. https://www.unicefusa.org/mission/emergencies/hurricanes/2013-philippines-typhoon-haiyan#.

15. Gutschow, B., B. Gray, M. I. Ragavan, P. E. Sheffield, R. P. Philipsborn, and S. H. Jee. "The Intersection of Pediatrics, Climate Change, and Structural Racism: Ensuring Health Equity Through Climate Justice." Curr Probl Pediatr Adolesc Health Care 51 (2021): 101028. [PMID: 34238692]
Frank, Thomas. "Population of Top 10 Counties for Disasters: 81% Minority." E&E News, 2020. https://www.eenews.net/articles/population-of-top-10-counties-for-disasters-81-minority.

16. Ibid.; Rysavy, T. F., and A. Floyd. "People of Color Are on the Front Lines of the Climate Crisis." Green America, 2016. https://www.greenamerica.org/clim ate-justice-all/people-color-are-front-lines-climate-crisis; Harvard Educational Review. "Katrina and Rita: What Can the United States Learn from International Experiences with Education in Displacement." Harvard Education Publishing Group, 2005. https://www.hepg.org/her-home/issues/harvard-educational-rev iew-volume-75-issue-4/herarticle/what-can-the-united-states-learn-from-inter nationa; Weems, C. F., L. K. Taylor, M. F. Cannon, R. C. Marino, D. M. Romano, B. G. Scott, A. M. Perry, and V. Triplett. "Post Traumatic Stress, Context, and the Lingering Effects of the Hurricane Katrina Disaster among Ethnic Minority Youth." J Abnormal Child Psychol 38 (2010): 49–56. https://doi.org/10.1007/s10 802-009-9352-y.

17. Wegienka, G., C. C. Johnson, E. Zoratti, and S. Havstad. "Racial Differences in Allergic Sensitization: Recent Findings and Future Directions." Curr Allergy Asthma Rep 13, 3 (2013): 255–261. https://doi.org/10.1007/s11882-013-0343-2; Ansorge, Rick. "Mold-Related Health Hazards Found in New Orleans Homes." HealthDay, 2006. https://consumer.healthday.com/allergy-1/safety-and-pub lic-health-news-585/mold-related-health-hazards-found-in-new-orleans-homes- 501371.html.

18. Charveriat, Céline. "The New European Parliament: Tackling the Triple Injustice of Climate Change." The Progressive Post, 2019. https://progressivepost.eu/ focus/new-european-parliament-tackling-triple-injustice-climate-change; Carmin, JoAnn, and Yan Zhang. "Achieving Urban Climate Adaptation in Europe and Central Asia." World Bank, 2009. http://documents1.worldbank.org/cura ted/en/744371468029956345/text/WPS5088.txt.

19. English, J. "Millions of Children Affected by Devastating Flooding in South Asia, with Many More at Risk as COVID-19 Brings Further Challenges." UNICEF, 2020. https://www.unicef.org/press-releases/millions-children-affected-devastat ing-flooding-south-asia-many-more-risk-covid-19.

20. World Wildlife Fund. "Is Climate Change Threatening the Saami Way of Life?" arcticwwf.org, 2021. https://arcticwwf.org/newsroom/the-circle/arctic-tipping- point/climate-change-culture-change/.

21. Herold, N., L. Alexander, D. Green, and M. Donat. "Greater Increases in Temperature Extremes in Low Versus High Income Countries." Environ Res Letters 12, 3 (2017): 1–4. https://doi.org/10.1088/1748-9326/aa5c43; Frank, "Population of Top 10"; Hoffman, J. S., V. Shandas, and N. Pendleton. "The Effects of Historical Housing Policies on Resident Exposure to Intra-Urban Heat: A Study of 108 US Urban Areas." Climate 8, 1 (2020): 12. https://doi.org/ 10.3390/cli8010012.

22. Plumer, Brad, and Nadja Popovich. "How Decades of Racist Housing Policy Left Neighborhoods Sweltering." New York Times, 2020. https://www.nytimes.com/ interactive/2020/08/24/climate/racism-redlining-cities-global-warming.html.

23. Geruso, M., and D. Spears. "Heat, Humidity, and Infant Mortality in the Developing World." National Bureau of Economic Research, 2018. https://www. nber.org/papers/w24870.

24. Sengupta, S. "Wildfire Smoke Is Poisoning California's Kids: Some Pay a Higher Price." New York Times, 2020. https://www.nytimes.com/interactive/2020/11/ 26/climate/california-smoke-children-health.html.

25. Davies, Ian P., Ryan D. Haugo, James C. Robertson, and Phillip S. Levin. "The Unequal Vulnerability of Communities of Color to Wildfire." PLoS ONE 13, 11

(2018): e0205825. https://doi.org/10.1371/journal.pone.0205825; Kim, L., M. Whitaker, A. O'Halloran, et al. "Hospitalization Rates and Characteristics of Children Aged <18 Years Hospitalized with Laboratory-Confirmed COVID-19: COVID-NET, 14 States, March 1–July 25, 2020." MMWR Morb Mortal Wkly Rep 69 (2020): 1081–1088. http://dx.doi.org/10.15585/mmwr.mm6932e3.

26. CAP Action. "Lessons on Climate Change and Poverty From the California Drought." August 19, 2015. https://www.americanprogress.org/article/lessons-on-climate-change-and-poverty-from-the-california-drought/

27. UNICEF. "Malnutrition." 2021. https://data.unicef.org/topic/nutrition/malnutrition/
    Global Nutrition Report. "Country Nutrition Profiles." 2021. https://globalnutritionreport.org/resources/nutrition-profiles/north-america/northern-america/united-states-america/
    U.S Department of Agriculture Economic Research Services. "Food Security in the U.S." 2022. https://www.ers.usda.gov/topics/food-nutrition-assistance/food-security-in-the-u-s/key-statistics-graphics/#children
    "U.S. by the Numbers." NPR, 2020. https://www.npr.org/2020/09/27/912486921/food-insecurity-in-the-u-s-by-the-numbers.

28. CDC. "Diarrhea: Common Illness, Global Killer." 12/7/12. https://www.cdc.gov/healthywater/pdf/global/programs/globaldiarrhea508c.pdf.

29. UNESCO. "New UNESCO Working Paper on the Impact of Climate Displacement on the Right to Education." 2020. https://www.weforum.org/agenda/2019/12/extreme-weather-climate-change-displaced/; Bassetti, F. "Environmental Migrants: Up to 1 Billion by 2050." Foresight, 2019. https://www.climateforesight.eu/migrations-inequalities/environmental-migrants-up-to-1-billion-by-2050/; Smith, H. "Shocking Images of Drowned Syrian Boy Show Tragic Plight of Refugees." The Guardian, 2015. https://www.theguardian.com/world/2015/sep/02/shocking-image-of-drowned-syrian-boy-shows-tragic-plight-of-refugees.

30. Pew Research Center. "Number of Refugees to Europe Surges to Record 1.3 Million in 2015." 2016. https://www.pewresearch.org/global/2016/08/02/number-of-refugees-to-europe-surges-to-record-1-3-million-in-2015/; Bierbach, M. "Migration to Europe in 2019: Facts and figures." Info Migrants, 2019. https://www.infomigrants.net/en/post/21811/migration-to-europe-in-2019-facts-and-figures; UNICEF. "Refugee and Migrant Children in Europe." 2020. https://www.unicef.org/eca/emergencies/refugee-and-migrant-children-europe.

31. Kardas-Nelson, M. "The Petrochemical Industry Is Killing Another Black Community in 'Cancer Alley'." The Nation, 2019. https://www.thenation.com/article/archive/st-james-louisiana-plastic-petrochemicals-buy-out/. Schleifstein, Mark. "I've Investigated Industrial Pollution for 35 Years. We're Going Backwards." ProPublica. October 30, 2019. https://www.propublica.org/article/ive-investigated-industrial-pollution-for-35-years-were-going-backwards

32. Varshavsky, J. R., S. Sen, J. F. Robinson, S. C. Smith, J. Frankenfield, Y. Wang, G. Yeh, et al. "Racial/Ethnic and Geographic Differences in Polybrominated Diphenyl Ether (PBDE) Levels Across Maternal, Placental, and Fetal Tissues During Mid-Gestation." Scientific Rep 10 (2020): 12247. https://doi.org/10.1038/s41598-020-69067-y; Ruiz, D., M. Becerra, J. S. Jagai, K. Ard, and R. M. Sargis. "Disparities in Environmental Exposures to Endocrine-Disrupting Chemicals and Diabetes Risk in Vulnerable Populations." Diabetes Care 41, 1 (2018): 193–205. https://doi.org/10.2337/dc16-2765.

33. Szalavitz, M. "How Stress Gets Under the Skin: Q&A With Neuroscientist Bruce McEwen." Time, 2013. https://healthland.time.com/2013/02/20/how-stress-gets-under-the-skin-qa-with-neuroscientist-bruce-mcewen/.

34. McCorry, L. K. "Physiology of the Autonomic Nervous System." Am J Pharm Educ 71, 4 (2007): 78. https://doi.org/10.5688/aj710478.

35. Harvard Health Publishing. "Understanding the Stress Response." July 6, 2020. https://www.health.harvard.edu/staying-healthy/understanding-the-stress-response

36. Burris, H. H., S. A. Lorch, H. Kirpalani, D. M. Pursley, M. A. Elovitz, and J. E. Clougherty. "Racial Disparities in Preterm Birth in USA: A Biosensor of Physical and Social Environmental Exposures." Archives of Disease in Childhood 104(10) (2019): 931–935. https://doi.org/10.1136/archdischild-2018-316486; Vishnevetsky, J., D. Tang, H. Chang, E. L. Roen, Y. Wang, V. Rauh, S. Wang, et al. "Combined Effects of Prenatal Polycyclic Aromatic Hydrocarbons and Material Hardship on Child IQ." Neurotoxicol Teratol 49 (2015): 74–80. https://doi.org/10.1016/j.ntt.2015.04.002; Perera, F. P., S. Wang, V. Rauh, H. Zhou, L. Stigter, D. Camann, W. Jedrychowski, E. Mroz, and R. Majewska. "Prenatal Exposure to Air Pollution, Maternal Psychological Distress, and Child Behavior." Pediatrics 132, 5 (2013): e1284–e1294. https://doi.org/10.1542/peds.2012-3844.

37. Van den Bergh, B., R. Dahnke, and M. Mennes. "Prenatal Stress and the Developing Brain: Risks for Neurodevelopmental Disorders." Developm Psychopathol 30, 3 (2018): 743–762. https://doi.org/10.1017/S0954579418000342.

38. Stoye, D. Q., M. Blesa, G. Sullivan, P. Galdi, G. J. Lamb, G. S. Black, A. J. Quigley, et al. "Maternal Cortisol Is Associated with Neonatal Amygdala Microstructure and Connectivity in a Sexually Dimorphic Manner." eLife 9 (2020): e60729. https://doi.org/10.7554/eLife.60729.

39. Luby, J., A. Belden, K. Botteron, et al. "The Effects of Poverty on Childhood Brain Development: The Mediating Effect of Caregiving and Stressful Life Events." JAMA Pediatr 167, 12 (2013): 1135–1142. https://doi.org/10.1001/jamapediatrics.2013.3139.

40. Luby, J., R. Tillman, and D. M. Barch. "Association of Timing of Adverse Childhood Experiences and Caregiver Support with Regionally Specific Brain Development in Adolescents." JAMA Netw Open 2, 9 (2019): e1911426. https://doi.org/10.1001/jamanetworkopen.2019.11426; Oshri, A., J. C. Gray, M. M. Owens, S. Liu, E. B. Duprey, L. H. Sweet, and J. MacKillop. "Adverse Childhood Experiences and Amygdalar Reduction: High-Resolution Segmentation Reveals Associations with Subnuclei and Psychiatric Outcomes." Child Maltreat 24, 4 (2019): 400–410. https://doi.org/10.1177/1077559519839491.

41. McEwen, Bruce S., Jason D. Gray, and Carla Nasca. "Recognizing Resilience: Learning from the Effects of Stress on the Brain." Neurobiology of Stress 1 (January 2015): 1–11. https://doi.org/10.1016/j.ynstr.2014.09.001

CHAPTER 5

1. Rawls, J. A Theory of Justice. Cambridge, MA: Belknap Press 1971; ; Miller, G. T., S. Spoolman, G. T. Miller, and S. Spoolman. "Environmental Science." United States: Brooks/Cole; 2012.

2. National People of Color Environmental Leadership Summit, Lee, C., & United Church of Christ. (1992). "The First National People of Color Environmental

Leadership Summit, the Washington Court on Capitol Hill." Washington, D.C., October 24-27, 1991: Proceedings. New York, N.Y: The Commission<<<REFC>>>.

3. World Commission on Environment and Development. (1987). "Our Common Future." Oxford: Oxford University Press.

4. Inside Climate News. "Q&A: A Pioneer of Environmental Justice Explains Why He Sees Reason for Optimism." 2020. https://insideclimatenews.org/news/18062020/robert-bullard-black environmental-justice network-relaunch/.

5. VOA. "Protests Support Floyd, Black Lives Matter on 3 Continents." 2020. https://www.voanews.com/a/usa nation-turmoil-george-floyd-protests prote sts-support-floyd-black-lives-matter-3-continents/6190652.html.

6. Bullard, R, Gardezi M., Chennault. C., and Dankbar H. "Interview with Dr. Robert Bullard." 2021. https://lib.dr.iastate.edu/cgi/viewcontent.cgi?article=1120&context=jctp.

7. McCorry, Jamie. "An Interview with Peggy Shepard of WE ACT for Environmental Justice." August 31, 2020. https://greenworkslending.com/we-act-for-environmental-justice/; Communication in 2019 from Cecil Corbin-Mark.

8. Klein, N., *This Changes Everything: Capitalism vs. the Climate*. London: Penguin, 2015. An infamous example is the Keystone XL pipeline proposed to carry bitumen.

9. Ibrahim, Hindou Oumarou. Ted Talks, 2020. https://www.ted.com/speakers/hin dou_oumarou_ibrahim.

10. Los Angeles Times. "How Faith Spotted Eagle Became the First Native American to Win an Electoral Vote for President." December 20, 2016. https://www.lati mes.com/nation/la-na-faithspotted-eagle-2016-story.html.

11. Homan, Jackie. "A Native American Activist Speaks Candidly About What It's Like." September 20, 2017. https://repeller.com/faith-spotted-eagle-activist/.

12. Johnson, Ayana Elizabeth, "All We Can Save: Truth, Courage, and Solutions for the Climate Crisis." *One World New York*, 2020 https://www.allwecansave.earth

13. Francis I. "Laudato Si'." *The Vatican*, 2015. https://www.vatican.va/content/francesco/en/encyclicals/documents/papa-francesco_20150524_enciclica-laud ato-si.html.

14. New York Times. "The Pope on the Climate." June 18, 2015. https://www.nyti mes.com/2015/06/18/opinion/the-pope-on-the-climate.html.

15. "Analysis* of Pope Francis' Encyclical Laudato Si." *Climate Feedback*, 2016.

16. The Guardian. "Angry US Republicans Tell Pope Francis to 'Stick with His Job and We'll Stick with Ours'." June 13, 2015. https://www.theguardian.com/us-news/2015/jun/13/climate-change-conservatives-catholic-teaching.

17. New York Times. "Pope Francis Aligns Himself with Mainstream Science on Climate." June 18, 2015. https://www.nytimes.com/2015/06/19/science/earth/pope-francis-aligns-himself-with-mainstream-science-on-climate.html.

18. Maza, Christina. "One Year Later, How a Pope's Message on Climate Has Resonated." June 24, 2016. https://www.csmonitor.com/Environment/2016/0624/One-year-later-how-a-Pope-s-message-on-climate-has-resonated.

19. Self-Determination. "The Impact of Laudato Si' on the Paris Climate Agreement." August 2018. https://lisd.princeton.edu/publications/impact-laudato-si'-paris-climate-agreement.

20. Inside Climate News. "Five Years After Speaking Out on Climate Change, Pope Francis Sounds an Urgent Alarm." August 7, 2020. https://insideclimatenews.org/news/07082020/climate-change-pope-francis/.

21. Crux Now. "Expert Calls the Science Behind the Papal Encyclical 'Watertight'." June 18, 2015. https://cruxnow.com/church/2015/06/expert-calls-the-science-behind-the-papal-Encyclical-watertight/; The Guardian. "The Pope's Encyclical on Climate Change: As It Happened." June 18, 2015. https://www.theguardian.com/environment/blog/live/2015/jun/18/pope-Encyclical-climate-change-live-react ion-analysis#block-5582b62de4b0c09f64bfa923.

22. Dalai Lama. *Ecology, Ethics, and Interdependence: The Dalai Lama in Conversation with Leading Thinkers on Climate Change.* John D. Dunne and Daniel Goleman (Eds.). Somerville, MA: Wisdom Publications, 2018.

23. Heuman, Linda. "Awakening from Climate Slumber." *Spring* 2019. https://www.lindaheuman.com/single-post/2019/04/26/now-on-the-news-stand-awakening-from-climate-slumber.

24. Ancient Dragon Zen Gate. "A Buddhist Declaration on Climate Change." March 14, 2015. https://www.ancientdragon.org/a-buddhist-declaration-on-climate-change/

25. Dalai Lama. "Op-Ed: Dalai Lama: It's Up to Us—and Especially Politicians—to Save Our Planet." *Los Angeles Times.* 2020. https://www.dalailama.com/messa ges/environment/its-up-to-us-and-especially-politicians-to-save-our-planet

26. Thurman, Robert. "The Meaning of the Dalai Lama for Today." *Lion's Roar,* 2003. https://www.lionsroar.com/the-meaning-of-the-dalai-lama-for-today/.

27. Mind and Life. "The Dalai Lama with Greta Thunberg and Leading Scientists: A Conversation on the Crisis of Climate Feedback Loops." September 10, 2021. https://www.mindandlife.org/event/the-dalai-lama-with-greta-thunberg-and-leading-scientists-a-conversation-on-the-crisis-of-climate-feedback-loops/.

28. Pew Research Center. "Why Muslims Are the World's Fastest-Growing Religious Group." April 6, 2017. https://www.pewresearch.org/fact-tank/2017/04/06/why-muslims-are-the-worlds-fastest-growing-religious-group/; Ancient Dragon Zen Gate.

29. Sustainable Business. "Islamic Leaders Issue Thoughtful, Strong Climate Declaration." 2015, https://www.sustainablebusiness.com/2015/08/islamic-lead ers-issue-thoughtful-strong-climate-declaration-53000/.

30. Aljazeera. "What Does Islam Say About Climate Change and Climate Action?" August 12, 2020. https://www.aljazeera.com/opinions/2020/8/12/what-does-islam-say-about-climate-change-and-climate-action.

31. The Conversation. "Young Muslim Women"; Bioneers. "Mishka Banuri: A First Generation Immigrant's Perspective on Youth Climate Justice." 2019. https://bioneers.org/mishka-banuri-first-generation-immigrant-youth-climate-justice-zstf1911/.

32. Yale Climate Connections. "Judaism and Climate Change." February 29, 2012. https://yaleclimateconnections.org/2012/02/judaism-and-climate-change/.

33. Arcworld. "Sustaining Our Vision: The Jewish Climate Change Campaign." 2021. http://www.arcworld.org/downloads/JewishClimateCampaign%20Dr aft%201.pdf.

34. Interfaith Center for Sustainable Development. "Jewish Climate Initiative." 2021. https://www.interfaithsustain.com/jewish-climate-initiative/.

35. IASS Potsdam. "Climate Action Takes Shape in Israel." December 8, 2019. https://www.iass-potsdam.de/en/news/climate-action-takes-shape-israel.

36. The Conversation. "Faith and Politics Mix to Drive Evangelical Christians' Climate Change Denial." September 9, 2020. https://theconversation.com/faith-and-politics-mix-to-drive-evangelical-christians-climate-change-denial-143

145; The Evangelical Climate Initiative. "Climate Change: An Evangelical Call to Action." *Influence Watch,* 2020. https://www.influencewatch.org/app/uploads/2020/08/climate-change-an-evangelical-call-to-action.-08.20.pdf.

37. DW News. "God and the Earth: Evangelical Take on Climate Change." July 3, 2019. https://www.dw.com/en/god-and-the-earth-evangelical-take-on-climate-change/a-47781433.

38. Interfaith Center for Sustainable Development. "In Africa." 2019. https://www.interfaithsustain.com/faith-inspired-renewable-energy-project-in-africa/.

39. Interfaith Mission Service. "Climate Crisis." 2013. https://www.interfaithmissionservice.org/social-justice/climate-crisis/interfaith/.

40. Intergovernmental Panel on Climate Change. "Fifth Assessment Report." 2014. https://www.ipcc.ch/assessment-report/ar5/.

41. Intergovernmental Panel on Climate Change. "Special Report: Global Warming of 1.5 °C." 2021. https://www.ipcc.ch/sr15/

42. Watts, N., et al. "The 2018 Report of the Lancet Countdown on Health and Climate Change: Shaping the Health of Nations for Centuries to Come." *Lancet 392*, 10163 (2018): 2479–2514.

43. Chakradhar, Shraddha. "New Pediatrician Network Puts Spotlight on Climate Change's Effects on Children." December 18, 2020. https://www.statnews.com/2020/12/18/new-pediatrician-network-puts-spotlight-on-climate-changes-effects-on-children/.

44. Ibid.; Georgia Bio. "New Pediatrician Network Puts Spotlight on Climate Change's Effects on Children." December 18, 2020. https://gabio.org/new-pediatrician-network-puts-spotlight-on-climate-changes-effects-on-children/.

45. Maibach, Edward, Jeni Miller, Fiona Armstrong, Omnia El Omrani, Ying Zhang, Nicky Philpott, Sue Atkinson, et al. "Health Professionals, the Paris Agreement, and the Fierce Urgency of Now." *J Climate Change Health 1* (Mar 2021): 100002. https://doi.org/10.1016/j.joclim.2020.100002.

46. CNN Health. "More Than 230 Journals Warn 1.5°C of Global Warming Could Be 'Catastrophic' for Health." September 5, 2020. https://www.cnn.com/2021/09/05/health/climate-health-journals-warning-intl/index.html.

47. Perera, F., and K. Nadeau. "Climate Change, Fossil-Fuel Pollution, and Children's Health." New England Journal of Medicine 386(24) (2022): 2303–2314. https://www.nejm.org/doi/full/10.1056/NEJMra2117706

48. Atlas of the Future. "Feel the Force: 1m Moms Fight for Cleaner Air." 2021. https://atlasofthefuture.org/project/moms-clean-air-force/.

49. 350.org. "Stop Fossil Fuels. Build 100% Renewables." 2021. https://350.org.

50. Gorman A. "Earthrise: A Poem by Amanda Gorman." 2019 https://www.youtube.com/watch?v=xwOvBv8RLmo.

51. Creative Boom. "John Akomfrah's Purple Looks at Climate Change and Its Effects on Our Planet." 2017. https://www.creativeboom.com/inspiration/john-akomfrah-purple/; Winkball Video. "John Akomfrah: Purple by Winkball." 2017. https://www.youtube.com/watch?v=Wp4FwxwjWl0.

52. Salas, R. N., W. Jacobs, and F. Perera, "The Case of Juliana v. U.S.: Children and the Health Burdens of Climate Change." *N Engl J Med 380*, 22 (2019): 2085–2087.

53. Klein, Naomi. "On Fire: The (Burning) Case for a Green New Deal." New York: Simon & Schuster, 2019 .

54. The Guardian, "Climate Crisis and a Betrayed Generation." March 1, 2019 https://www.theguardian.com/environment/2019/mar/01/youth-climate-cha nge-strikers-open-letter-to-world-leaders.
55. Democracy Now. "'We Are Striking to Disrupt the System': *An Hour with 16-Year-Old Climate Activist Greta Thunberg.* September 11, 2019 https://innerself.com/ climateimpactnews/politics/3346-we-are-striking-to-disrupt-the-system-an-hour-with-16-year-old-climate-activist-greta-thunberg
56. Democracy Now. "The Marshall Islands Are Drowning." December 12, 2019. https://www.youtube.com/watch?v=hdp26q3Yseg.
57. Yes Magazine. "There Is No Climate Justice Without Racial Justice." June 7, 2020. https://www.yesmagazine.org/environment/2020/06/12/climate-justice-racial-justice; The Action Network. "This Is Zero Hour." 2021. https://actionnetw ork.org/groups/youth-march-for-climate-action-now?source=widget; Margolin, Jamie. "Jamie Margolin's 2019 Congressional Testimony." 2019. https://www. congress.gov/116/meeting/house/109951/witnesses/HHRG-116-FA14-Wstate-MargolinJ-20190918.pdf; Margolin, Jamie. "Youth to Power: Your Voice and How to Use It." 2016. https://www.hachettebookgroup.com/titles/jamie-margo lin/youth-to-power/9780738246666/.
58. Extinction Rebellion. 2021. https://rebellion.global/why-rebel/.
59. Yale Climate Connections. "The Society of Fearless Grandmothers Stands Up for Young Climate Activists." June 22, 2020. https://yaleclimateconnections.org/ 2020/06/the-society-of-fearless-grandmothers-stands-up-for-young-climate-activists/?ct=t(EMAIL_CAMPAIGN_WEEKLY_06222).
60. Rolling Stone. "Jane Fonda Fights On." March 31, 2020. https://www.rollingst one.com/culture/culture-features/jane-fonda-fire-drill-fridays-965715/.
61. Wikipedia. "The Elders." 2021. https://en.wikipedia.org/wiki/The_Elders_(organ ization)#Climate_change.
62. Dalai Lama, "It's Up to Us"; The Shalom Center. "Elijah's Covenant." *New Rabbinic Statement on the Climate Crisis.* 2020. https://theshalomcenter.org/cont ent/elijahs-covenant-new-rabbinic-statement-climate-crisis.
63. Ibid.

CHAPTER 6

1. Markandya, A., J. Sampedro, S. J. Smith, R. Van Dingenen, C. Pizarro-Irizar, I. Arto, and M. González-Eguino. "Health Co-Benefits from Air Pollution and Mitigation Costs of the Paris Agreement: A Modelling Study." *Lancet Planetary Health* 2, 3 (2018): e126–1e33. https://doi.org/10.1016/S2542-5196(18)30029-9.
2. European Commission. "EU Emissions Trading System (EU ETS)." 2021. https:// ec.europa.eu/clima/policies/ets_en.
3. Holland, M. R. *"The Co-Benefits to Health of a Strong EU Climate Change Policy."* World Wildlife Foundation EU, September 2008. https://wwfeu.awsassets. panda.org/downloads/co_benefits_to_health_report__september_2008. pdf; Liboreiro, Jorge. "Why Is the EU's new Emissions Trading System so Controversial?" August 2021. https://www.euronews.com/2021/07/16/why-is-the-eu-s-new-emissions-trading-system-so-controversial.
4. Huang, J., X. Pan, X. Guo, and G. Li. "Health Impact of China's Air Pollution Prevention and Control Action Plan: An Analysis of National Air Quality Monitoring and Mortality Data." *Lancet Planetary Health* 2, 7 (July 1 2018): e313–e323. https://doi.org/10.1016/S2542-5196(18)30141-4.

5. Liu, J., R. T. Woodward, and Y. Zhang. "Has Carbon Emissions Trading Reduced PM2.5 in China?" *Environmental Science & Technology 55*, 10 (May 7 2021): 6631–6643. https://doi.org/10.1021/acs.est.1c00248.
6. International Carbon Action Partnership. "China National ETS." 2021. https://icapcarbonaction.com/en/ets/china-national-ets; Chang., S., X. Yang, H. Zheng, S. Wang, and X. Zhang. "Air Quality and Health Co-Benefits of China's National Emission Trading System." *Applied Energy 261* (March 1 2020): 114226. https://doi.org/10.1016/j.apenergy.2019.114226.
7 US EPA Office of Air and Radiation. "The Benefits and Costs of the Clean Air Act from 1990 to 2020. March 2011. https://www.epa.gov/sites/production/files/2015-07/documents/summaryreport.pdf.
8. Simeonova, E., J. Currie, P. Nilsson, and R. Walker. "Congestion Pricing, Air Pollution, and Children's Health." *J Human Resources 56*, 4 (October 14, 2019): 971–996 . https://doi.org/10.3368/jhr.56.4.0218-9363R2.
9. Blue Sky Analytics. "Air Pollution History: The Great Smog of London." September 22, 2020. https://blueskyhq.in/blog/air-pollution-history-the-great-smog-of-london.
10. Santos, Georgina. "London Congestion Charging." *Brookings-Wharton Papers on Urban Affairs* (January 1, 2008): 177–234. https://www.jstor.org/stable/25609 551; Miller, Brian G. "Report on Estimation of Mortality Impacts of Particulate Air Pollution in London." June 2010. https://www.aef.org.uk/uploads/IomRepor t_1.pdf.
11. Kings College London. "Understanding the Health Impacts of Air Pollution in London." 2015. https://data.london.gov.uk/dataset/understanding-health-impa cts-of-air-pollution-in-london-.
12. Walton, Heather, David Dajnak, Dimitris Evangelopoulos, and Daniela Fecht. "*Health Impact Assessment of Air Pollution on Asthma in London.*" King's College London, April 2019. http://erg.ic.ac.uk/Research/home/Health%20Impact%20 Assessment%20Of%20Air%20Pollution%20On%20Asthma%20In%20London. pdf; Environmental Research Group King's College London. "Personalising The Health Impacts Of Air Pollution: Summary for Decision Makers." November 2019. http://erg.ic.ac.uk/Research/home/projects/personalised-health-impa cts.html.
13. Vaughan, Adam. "Boris Johnson Accused of Burying Study Linking Pollution and Deprived Schools." *The Guardian*, May 2016. https://www.theguardian.com/envi ronment/2016/may/16/boris-johnson-accused-of-burying-study-linking-pollut ion-and-deprived-schools.
14. Greater London Authority. "ULEZ reduces 13,500 cars daily & cuts toxic air pollution by a third." London Assembly, October 21, 2019. https://www.london. gov.uk/press-releases/mayoral/ulez-reduces-polluting-cars-by-13500-every-day.
15. Ibid.; London Assembly. "World's First Ultra Low Emission Zone to Save NHS Billions by 2050." February 2020. https://www.london.gov.uk/press-releases/ mayoral/ulez-to-save-billions-for-nhs.
16. Mayor of London. "Air Quality in London 2016–2020." 2020. https://www.lon don.gov.uk/what-we-do/environment/pollution-and-air-quality/air-quality-lon don-2016-2020.
17. Greater London Authority. "London Environment Strategy." May 2018. https:// www.london.gov.uk/sites/default/files/london_environment_strategy_0.pdf.
18. The Guardian. "'Dramatic' Plunge in London Air Pollution Since 2016, Report Finds." October 3, 2020. https://www.theguardian.com/environment/2020/oct/

03/dramatic-plunge-in-london-air-pollution-since-2016-report-finds; Dr. Gary Fuller is quoted in Mayor of London, "Air Quality in London."

19. The Guardian. "'Dramatic' Plunge"

20. London Assembly. "92 Per Cent of Vehicles Comply with Expanded ULEZ One Month On." *Greater London Authority*, December 2021. https://www.london.gov.uk/press-releases/mayoral/92-per-cent-of-vehicles-comply-with-expanded-ulez

21. Greater London Authority. "London Environment Strategy."

22. World Resources Institute. "5 Finalists for 2020-21 Prize for Cities Show How to Tackle Climate Change and Inequality Together." December 16, 2020. https://www.wri.org/news/release-5-finalists-2020-21-prize-cities-show-how-tackle-climate-change-and-inequality.

23. "Smog over Krakow, Poland." 2017. https://www.shutterstock.com/image-photo/cracow-poland-january-29-2017-smog-751991392.

24. Perera, F. P., W. Jedrychowski, V. Rauh, and R. M. Whyatt. "Molecular Epidemiologic Research on the Effects of Environmental Pollutants on the Fetus." *Environ Health Perspect 107*, 3 (1999): 451–460. https://doi.org/10.1289/ehp.99107s3451.

25. Edwards, S. C., W. Jedrychowski, M. Butscher, D. Camann, A. Kieltyka, E. Mroz, E. Flak, et al. "Prenatal Exposure to Airborne Polycyclic Aromatic Hydrocarbons and Children's Intelligence at 5 Years of Age in a Prospective Cohort Study in Poland." *Research Support, N.I.H., Extramural. Environ Health Perspect 118*, 9 (September 2010): 1326–1331. https://doi.org/10.1289/ehp.0901070; Jedrychowski, W. A., F. Perera, D. Camann, J. Spengler, M. Butscher, E. Mroz, R. Majewska, et al. "Prenatal Exposure to Polycyclic Aromatic Hydrocarbons and Cognitive Dysfunction in Children." *Environ Sci Pollut Res Int 22*, 5 (March 2015): 3631–3639. https://doi.org/10.1007/s11356-014-3627-8; Perera, F. P., S. Wang, V. Rauh, H. Zhou, L. Stigter, D. Camann, W. Jedrychowski, E. Mroz, and R. Majewska. "Prenatal Exposure to Air Pollution, Maternal Psychological Distress, and Child Behavior." *Pediatrics 132*, 5 (November 2013): e1284–1294. https://doi.org/10.1542/peds.2012-3844; Jedrychowski, W., A. Galas, A. Pac, E. Flak, D. Camman, V. Rauh, and F. Perera. "Prenatal Ambient Air Exposure to Polycyclic Aromatic Hydrocarbons and the Occurrence of Respiratory Symptoms over the First Year of Life." *Eur J Epidemiol 20*, 9 (2005): 775–782. https://doi.org/10.1007/s10654-005-1048-1; Jedrychowski, W., F. Perera, U. Maugeri, J. D. Spengler, E. Mroz, V. Rauh, E. Flak, et al. "Effect of Prenatal Exposure to Fine Particles and Postnatal Indoor Air Quality on the Occurrence of Respiratory Symptoms in the First Two Years of Life." *Int J Environ Health 2*, 3-4 (2008): 314–329. https://doi.org/10.1504/ijenvh.2008.020925; Jedrychowski, W., U. Maugeri, E. Mroz, E. Flak, M. Rembiasz, R. Jacek, and A. Sowa. "Fractional Exhaled Nitric Oxide in Healthy Non-Asthmatic 7-Year Olds and Prenatal Exposure to Polycyclic Aromatic Hydrocarbons: Nested Regression Analysis." *Pediatr Pulmonol* (May 15, 2012). https://doi.org/10.1002/ppul.22570; Jedrychowski, W. A., F. Perera, R. Majewska, D. Camman, J. D. Spengler, E. Mroz, L. Stigter, E. Flak, and R. Jacek. "Separate and Joint Effects of Tranplacental and Postnatal Inhalatory Exposure to Polycyclic Aromatic Hydrocarbons: Prospective Birth Cohort Study on Wheezing Events." *Pediatr Pulmonol 49*, 2 (February 2014): 162–172. https://doi.org/10.1002/ppul.22923; Jedrychowski, W. A., F. Perera, U. Maugeri, E. Mroz, M. Klimaszewska-Rembiasz, E. Flak, S. Edwards, and J. D. Spengler. "Effect of Prenatal Exposure to Fine Particulate Matter on Ventilatory Lung Function of Preschool Children of Non-Smoking Mothers."

Perinat Epidemiol 24, 5 (September 2010): 492–501. https://doi.org/10.1111/ j.1365-3016.2010.01136.x; Majewska, R., A. Pac, E. Mróz, J. D. Spengler, D. E. Camann, D. Mrozek-Budzyn, A. Sowa, et al. "Lung Function Growth Trajectories in Non-Asthmatic Children Aged 4-9 in Relation to Prenatal Exposure to Airborne Particulate Matter in Polycyclic Aromatic Hydrocarbons: Krakow Birth Cohort Study." *Environ Res 166* (2018): 150–157. https://doi.org/10.1016/ j.envres.2018.05.037; Jedrychowski, W. A., F. Perera, J. D. Spengler, E. Mroz, L. Stigter, E. Flak, R. Majewska, M. Klimaszewska-Rembiasz, and R. Jacek. "Intrauterine Exposure to Fine Particulate Matter As a Risk Factor for Increased Susceptibility to Acute Broncho-Pulmonary Infections in Early Childhood." *Int J Hyg Environ Health 216*, 4 (July 2013): 395–401. https://doi.org/10.1016/ j.ijheh.2012.12.014.

26. Lochno, A. "Expert Report on Possibility of Introducing Restriction of Solid Fuels Usage in Area of Krakow." *[In Polish.]* (2011); Junninen, H., J. Monster, M. Rey, et al. "Quantifying the Impact of Residential Heating on the Urban Air Quality in a Typical European Coal Combustion Region." *Environ Sci Technol 43*, 20 (October 2009): 7964–7960.

27. Majewska, R., A. Sowa, R. Jacek, K. Piotrowicz, F P Perera, J D Spengler, D E Camann, and A Pac. "Changes in the Air Pollution Levels During Heating Seasons in Krakow Based on Individual Measurements of Pollutants over the Last 15 Years." *Environ Epidemiol 3*, 257 (October 2019). https://doi.org/10.1097/ 01.EE9.0000608724.51778.a6.

28. Coal stoves were largely removed by 2019, prior to the ban on burning of solid fuels in Krakow stoves, effective September 1, 2019. We did not include the levels for 2020 since air quality was strongly influenced by COVID-19.

29. Gomez, J. M., and D. Bartyzel. "Drones Target Polluters in One of Europe's Smoggiest Places." *Bloomberk Businessweek,* January 15, 2020. https://www. bloomberg.com/news/features/2020-01-15/krakow-is-policing-clean-air-with- drones.

30. Climate-KIC. "Krakow: Transforming the City Towards Climate Neutrality." April 28, 2021. https://www.climate-kic.org/news/krakow-transforming-the-city- towards-climate-neutrality/; Krakow Post. "Young People March Against Climate Change in Krakow." March 17, 2019. http://www.krakowpost.com/20100/2019/ 03/school-strike-for-climate-krakow.

31. Kaufman, N. "Carbon Tax vs. Cap-and-Trade: What's a Better Policy to Cut Emissions?" *World Resources Institute,* March 1, 2016. https://www.wri.org/insig hts/carbon-tax-vs-cap-and-trade-whats-better-policy-cut-emissions.

32. RGGI Inc. "The Regional Greenhouse Gas Initiative: An Initiative of the New England and Mid-Atlantic States of the US." 2021. https://www.rggi.org/; Center for Climate and Energy Solutions. "Regional Greenhouse Gas Initiative (RGGI)." (2019).

33. RGGI Inc. "The Regional Greenhouse Gas."

34. Hodan, William M, and William R. Barnard. "Evaluating the Contribution of PM2.5 Precursor Gases and Re-entrained Road Emissions to Mobile Source PM2.5 Particulate Matter Emissions." *MACTEC* (epa.gov). https://www3.epa. gov/ttnchie1/conference/ei13/mobile/hodan.pdf.

35. Abt Associates. "Analysis of the Public Health Impacts of the Regional Greenhouse Gas Initiative, 2009–2014." 2017. https://www.abtassociates.com/ insights/publications/report/analysis-of-the-public-health-impacts-of-the-regio nal-greenhouse-gas.

Perera, F., D. Cooley, A. Berberian, D. Mills, and P. Kinney. "Co-Benefits to Children's Health of the U.S. Regional Greenhouse Gas Initiative." *Environ Health Perspect 128*, 7 (2020). https://doi.org/10.1289/EHP6706;

36. Acadia Center. "The Regional Greenhouse Gas Initiative: 10 Years in Review." 2019. https://acadiacenter.org/wp-content/uploads/2019/09/Acadia-Cen ter_RGGI_10-Years-in-Review_2019-09-17.pdf.

37. Acadia Center. "The Regional Greenhouse Gas."

38. RGGI Inc. "The Regional Greenhouse Gas"; RGGI Inc. "RGGI States Release Fourth Control Period Compliance Report." RGGI CO2 Allowance Tracking System, 2021. https://www.rggi.org/sites/default/files/Uploads/Press-Releases/ 2021_04_02_FoCP_Compliance.pdf.

39. RGGI Inc. "RGGI States Announce Proposed Program Changes: Additional 30% Emissions Cap Decline by 2030." RGGI Inc, August 23, 2017, https://www.rggi. org/sites/default/files/Uploads/Program-Review/8-23-2017/Announcement_P roposed_Program_Changes.pdf.

40. McKeown, Megan. "Carbon Trading & Environmental Equity: Evidence from the Regional Greenhouse Gas Initiative (2000–2019)." *Master's dissertation, University of Washington*, 2020. https://digital.lib.washington.edu/researchwo rks/handle/1773/46833?show=full.

41. Mendez, M. *Climate Change from the Streets: How Conflict and Collaboration Strengthen the Environmental Justice Movement.* New Haven, CT: Yale University Press, 2020.

42. California Air Resources Board. "Local Government Actions for Climate Change." California Air Resources Board, 2022. https://ww2.arb.ca.gov/our-work/progr ams/local-actions-climate-change/local-government-actions-climate-change.

43. California Climate Investments. 2021. "Annual Report: Cap-and-Trade Auction Proceeds." ca.gov, 2021. https://ww2.arb.ca.gov/sites/default/files/classic/cc/ capandtrade/auctionproceeds/2021_cci_annual_report.pdf.

44. O'Connor, T., K. Hsia-Kiung, L. Koehler, B. Holmes-Gen, W. Barrett, M. Chan, and K. Law. "Driving California Forward Public Health and Societal Economic Benefits of California's AB 32 Transportation Fuel Policies." Environmental Defense Fund, 2014. https://www.edf.org/sites/default/files/content/edf_ driving_california_forward.pdf; Breslow, M., and R. Wincele. "Cap-and-Trade in California: Health and Climate Benefits Greatly Outweigh Costs." Climate XChange, March 2020. https://climate-xchange.org/wp-content/uploads/2018/ 08/California_Cap_and_Trade-3-13-2020-spreads.pdf.

45. Mendez, "Climate Change from the Streets."

46. Cushing, L., D. Blaustein-Rejto, M. Wander, M. Pastor, J. Sadd, A. Zhu, and R. Morello-Frosch. "Carbon Trading, Co-Pollutants, and Environmental Equity: Evidence from California's Cap-and-Trade Program (2011–2015)." *PLoS Medicine 15*, 7 (July 10, 2018): e1002604. https://doi.org/10.1371/journal. pmed.1002604; California Climate Investments. 2021 "Annual Report; California Air Resources Board. FAQ Cap-and-Trade Program." *California Air Resources Board*, 2021. https://ww2.arb.ca.gov/resources/documents/faq-cap-and-trade- program.

47. Mendez, "Climate Change from the Streets."

48. Johnson, S., J. Haney, L. Cairone, C. Huskey, and I. Kheirbek. "Assessing Air Quality and Public Health Benefits of New York City's Climate Action Plans." *Environ Sci Technol 54* (2020): 9804–9813. https://doi.org/10.1021/acs.est.0c00 694; Kheirbek, Iyad, Katherine Wheeler, Sarah Walters, Grant Pezeshki, and

Daniel Kass. "Air Pollution and the Health of New Yorkers: The Impact of Fine Particles and Ozone." NYC Health. https://www1.nyc.gov/assets/doh/downlo ads/pdf/eode/eode-air-quality-impact.pdf; Pinto de Moura, M. C., D. Reichmuth, and D. Gatti. "Inequitable Exposure to Air Pollution from Vehicles in New York State." *Union of Concerned Scientists*, 2019. https://www.ucsusa.org/sites/default/ files/attach/2019/06/Inequitable-Exposure-to-Vehicle-Pollution-NY.pdf.

49. New York City Panel on Climate Change. 2019. Report Executive Summary. (Annals of the New York Academy of Sciences: 2019); Petkova, E. P., J. K. Vink, R. M. Horton, A. Gasparrini, D. A. Bader, J. D. Francis, and P. L. Kinney. "Towards More Comprehensive Projections of Urban Heat-Related Mortality: Estimates for New York City under Multiple Population, Adaptation, and Climate Scenarios." *Environmental Health Perspectives 125*, 1 (2017). https:// doi.org/10.1289/EHP166; Rosenzweig, C., and W. Solecki. "Hurricane Sandy and Adaptation Pathways in New York: Lessons from a First-Responder City." *Global Environ Change 28* (2014): 395–408. https://doi.org/10.1016/j.gloenv cha.2014.05.003.

50. Fry, D., M. Kioumourtzoglou, C. A. Treat, K. R. Burke, D. Evans, L. P. Tabb, D. Carrion, F. P. Perera, and G. S. Lovasi. "Development and Validation of a Method to Quantify Benefits of Clean-Air Taxi Legislation." *J Expo Sci Environ Epidemiol 30*, 4 (2020): 629–640. https://doi.org/10.1038/s41370-019-0141-6.

51. Carrión, D., W. V. Lee, and D. Hernández. "Residual Inequity: Assessing the Unintended Consequences of New York City's Clean Heat Transition." *Int J Environ Res Public Health 15*, 1 (2018): 117. https://doi.org/10.3390/ijerph15010 117; Zhang, L., M. Z. He, E. A. Gibson, F. P. Perera, G. S. Lovasi, J. Clougherty, D. Carrión, et al. "Evaluating the Impact of the Clean Heat Program on Air Pollution Levels in New York City." *Environ Health Perspectives.* 2021. https://doi.org/ 10.1289/EHP9976

52. Kheirbek, I., J. Haney, S. Douglas, K. Ito, S. Caputo Jr., and T. Matte. "The Public Health Benefits of Reducing Fine Particulate Matter through Conversion to Cleaner Heating Fuels in New York City." *Environ Sci Technol 48* (2014): 13573– 13582. dx.doi.org/10.1021/es503587p.

53. Perera, F. P., A. Berberian, D. Cooley, E. Shenaut, H. Olmstead, Z. Ross, and T. Matte. "Potential Health Benefits of Sustained Air Quality Improvements in New York City: A Simulation Based on Air Pollution Levels During the COVID-19 Shutdown." *Environ Res 193* (2021): 100555. https://doi.org/10.1016/j.env res.2020.110555.

CHAPTER 7

1. This chapter is intended to provide an overview of the various solutions available, each one deserving of a book in itself. The author hopes that it will be a helpful guide to readers who are less familiar with the topic.

2. UNICEF. "The Climate Crisis Is a Child Rights Crisis: Introducing the Children's Climate Risk Index." 2021. https://www.unicef.org/media/105376/file/UNICEF-

3. Cabot Venton, C. "The Benefits of a Child-Centered Approach to Climate Change Adaptation." UNICEF; Plan International. August 2021. https://www.uncclearn. org/wp-content/uploads/library/unicef02.pdf.

4. CarbonBrief. "Climate Strikers: Open Letter to EU Leaders on Why Their New Climate Law Is 'Surrender'." CarbonBrief: Clear on Climate, March 3 2020. https://www.carbonbrief.org/climate-strikers-open-letter-to-eu-leaders-on-why-their-new-climate-law-is-surrender.

5. Gormon, Amanda. "Earthrise." January 20 2021. Poem. Performed live at the Inauguration of President Joe Biden and Vice President Kamala Harris in Washington D.C.

6. Major sources for this chapter are the excellent analyses by Project Drawdown from which much of this information is drawn: Foley, J. "We Need Four Waves of Climate Action." 2021. https://www.drawdown.org/news/insights/we-need-four-waves-of-climate-action; "The Drawdown Review." 2021. https://drawdown.org/drawdown-review. I have followed the outline in those excellent publications. Here, it is only possible to give an overview of this very complex topic, providing examples and references so that the reader may take a deeper dive.

7. World Resources Institute. "5 Strategies That Achieve Climate Mitigation and Adaptation Simultaneously." February 10, 2020. https://www.wri.org/insights/5-strategies-achieve-climate-mitigation-and-adaptation-simultaneously.

8. Lamb, William F. "What Are the Social Outcomes of Climate Policies? A Systematic Map and Review of the Ex-Post Literature." Environ Res Letters 15 (2020): 113006.

9. Project Drawdown, "The Drawdown Review."

10. International Renewable Energy Agency (IRENA). "Renewable Capacity Statistics." 2021. https://www.irena.org/publications/2021/March/Renewable-Capacity-Statistics-2021; IEA. "Renewables 2021 Analysis and Forecasts to 2026." December 2021. https://www.iea.org/reports/renewables-2021.

11. U.S. Department of Energy. "LED Lighting. " 2021. https://www.energy.gov/energysaver/save-electricity-and-fuel/lighting-choices-save-you-money/led-lighting; EarthTronics. "Earth Day 2020: Benefits of LED Lighting." 2020. https://www.earthtronics.com/2020/04/22/earth-day-2020-benefits-of-led-lighting/.

12. WeAct For Environmental Justice. "Solar Uptown Now Shines Light on a Just Transition to Clean Energy and Green Jobs in Harlem." December 5, 2018. https://www.weact.org/2018/12/solar-uptown-now-shines-light-on-a-just-transition-to-clean-energy-and-green-jobs-in-harlem/.

13. Project Drawdown, "The Drawdown Review."

14. IEA. "Transport Improving the Sustainability of Passenger and Freight Transport." 2021. https://www.iea.org/topics/transport; World Resources Institute. "Electric School Bus Initiative." 2022. https://www.wri.org/initiatives/electric-school-bus-initiative.

15. Institute for Transportation & Development Policy. "Pedestrians First." 2021. https://pedestriansfirst.itdp.org.

16. Global Street Design Guide. "Pedestrian Only Streets: Case Study: Stroget, Copenhagen." Island Press. 2018. https://globaldesigningcities.org/publication/global-street-design-guide/streets/pedestrian-priority-spaces/pedestrian-only-streets/pedestrian-streets-case-study-stroget-copenhagen/; City Planning Institute of Japan. "The Human Dimension: A Sustainable Approach to City Planning." 2021. https://www.cpij.or.jp/eng/file/gehl.pdf; Peters, Adele. "In This Danish City, 5-Year-Olds Bike to School on Their Own." 2016. https://www.fastcompany.com/3057379/in-this-danish-city-5-year-olds-bike-to-school-on-their-own.

17. Foley, "We Need Four Waves."

18. Balakrishnan, Kalpana, Thomas Clasen, Sumi Mehta, Jennifer Peel, Ajay Pillarisetti, Amod Pokhrel, Jonathan Samet, Lisa Thompson, and Junfeng Zhang. "In Memoriam: Kirk R. Smith." Environ Health Perspect 128, 7 (2020): 071601. https://doi.org/doi:10.1289/EHP7808.

19. Lipinski, Brian. "By the Numbers: Reducing Food Loss and Waste." World Resources Institute, June 2013. https://www.wri.org/insights/numbers-reducing-food-loss-and-waste.

20. BEUC. "One Bite at a Time: Consumers and the Transition to Sustainable Food." June 2020. https://www.beuc.eu/publications/beuc-x-2020-042_consumers_and_the_transition_to_sustainable_food.pdf; Blaustein-Rejto, Dan, and Alex Smith. "We're on Track to Set a New Record for Global Meat Consumption." MIT Technology Review, April 2021. https://www.technologyreview.com/2021/04/26/1023636/sustainable-meat-livestock-production-climate-change/.

21. Foley, "We Need Four Waves."

22. BBC Future Planet. "The Batteries That Could Make Fossil Fuels Obsolete." December 17, 2020. https://www.bbc.com/future/article/20201217-renewable-power-the-worlds-largest-battery; U.S. Department of Energy. "Secretary Granholm Announces New Goal to Cut Costs of Long Duration Energy Storage by 90 Percent." July 14, 2021. https://www.energy.gov/articles/secretary-granholm-announces-new-goal-cut-costs-long-duration-energy-storage-90-percent; The White House. "Fact Sheet: The Bipartisan Infrastructure Deal." 2021. https://www.whitehouse.gov/briefing-room/statements-releases/2021/11/06/fact-sheet-the-bipartisan-infrastructure-deal/.

23. foodtank. "17 Organizations Promoting Regenerative Agriculture Around the Globe." May 2018. https://foodtank.com/news/2018/05/organizations-feeding-healing-world-regenerative-agriculture-2/; Matthews, Alan. "Agriculture in the European Green Deal: From Ambition to Action." Cap Reform, October 21, 2020. http://capreform.eu/agriculture-in-the-european-green-deal-from-ambition-to-action/

24. Wickenden, Dorothy. "Wendell Berry's Advice for a Cataclysmic Age." The New Yorker, February 21, 2022. https://www.newyorker.com/magazine/2022/02/28/wendell-berrys-advice-for-a-cataclysmic-age.

25. IUCN. "Peatlands and Climate Change." 2021. https://www.iucn.org/resources/issues-briefs/peatlands-and-climate-change.

26. Baragwanath, Kathryn, and Ella Bayi. "Collective Property Rights Reduce Deforestation in the Brazilian Amazon." Proc Nat Acad Sci 117, 34 (2020): 20495–20502. https://doi.org/10.1073/pnas.1917874117; Conservation International. "What on Earth Is 'Land Tenure'?" October 6, 2016. https://www.conservation.org/blog/what-on-earth-is-land-tenure; World Resources Institute. "Climate Benefits Tenure Costs." 2016. https://files.wri.org/d8/s3fs-public/Climate_Benefits_Tenure_Costs_Executive_Summary.pdf; World Wildlife Fund. "Deforestation and Forest Degradation." 2021. https://www.worldwildlife.org/threats/deforestation-and-forest-degradation.

27. PEW. "Ocean Protections Increasingly Seen as Key to Countering Climate Change." November 30, 2020. https://www.pewtrusts.org/en/research-and-analysis/articles/2020/12/01/ocean-protections-increasingly-seen-as-key-to-countering-climate-change.

28. World Wildlife. "Saving Mangroves." Summer 2018. https://www.worldwildlife.org/magazine/issues/summer-2018/articles/saving-mangroves.

29. Greenfield, Patrick, and Fiona Harvey. "More Than 50 Countries Commit to Protection of 30% of Earth's Land and Oceans." The Guardian, January 2021. https://www.theguardian.com/environment/2021/jan/11/50-countries-commit-to-protection-of-30-of-earths-land-and-oceans.

30. Foley, "We Need Four Waves."

31. Gertner, Jon. "Carbon Capture—Dream Or Nightmare—Could Be Coming. Or Not." Bull Atomic Scientists, August 2021. https://thebulletin.org/2021/08/carbon-capture-dream-or-nightmare-could-be-coming-or-not/; Project Drawdown, "The Drawdown Review"; Foley, "We Need Four Waves."

32. Tabuchi, Hiroko. "For Many, Hydrogen Is the Fuel of the Future: New Research Raises Doubts." New York Times, August 12, 2021. https://www.nytimes.com/2021/08/12/climate/hydrogen-fuel-natural-gas-pollution.html.

33. Service, Robert F. "Can the World Make the Chemicals It Needs Without Oil?" Science, September 2019. https://www.science.org/news/2019/09/can-world-make-chemicals-it-needs-without-oil; Zhao, Weijie. "Make the Chemical Industry Clean with Green Chemistry: An Interview with Buxing Han." Nat Sci Rev 5, 6 (April 2018): 953–956. https://doi.org/10.1093/nsr/nwy045.

34. Breakthrough. "Culture Change: A Breakthrough Perspective." Breakthrough, 2017 https://us.breakthrough.tv/resources/culture-change-breakthrough-perspective/

35. Madison Square Park Conservancy. "Maya Lin: Ghost Forest." Madison Square Park Conservancy, 2021, https://madisonsquarepark.org/art/exhibitions/maya-lin-ghost-forest/; Small, Zachary. "Maya Lin's Dismantled 'Ghost Forest' to be Reborn as Boats." The New York Times, November 24, 2021, https://www.nytimes.com/2021/11/24/arts/design/maya-lin-rocking-the-boat.html

36. United Nations. "Press Conference on 'Paint For The Planet' Exhibition." October 23, 2008. https://www.un.org/press/en/2008/081023_UNEP.doc.htm.

37. Guterres, António. "Tackling Inequality: A New Social Contract for a New Era." United Nations, 2020. https://www.un.org/en/coronavirus/tackling-inequality-new-social-contract-new-era.

38. Garfinkel, Irwin, Laurel Sariscsany, Elizabeth Ananat, Sophie Collyer, and Christopher Wimer. "The Costs and Benefits of a Child Allowance." Poverty Social Policy Brief 5, 1 (August 2021).

39. United Nations. "Paris Climate Deal Could Go Up in smoke Without Action: Guterres." September 2021. https://news.un.org/en/story/2021/09/1100242.

40. C40 Cities. "Cities Race to Zero." C40 Cities, 2021, https://www.c40.org/what-we-do/building-a-movement/cities-race-to-zero/.

41. Environmental and Energy Study Institute. "Fact Sheet | Fossil Fuel Subsidies: A Closer Look at Tax Breaks and Societal Costs." July 29, 2021. https://www.eesi.org/papers/view/fact-sheet-fossil-fuel-subsidies-a-closer-look-at-tax-breaks-and-societal-costs.

42. Carlin, David. "The Case for Fossil Fuel Divestment." Forbes, February 2021. https://www.forbes.com/sites/davidcarlin/2021/02/20/the-case-for-fossil-fuel-divestment/?sh=2f78003a76d2; Ambrose, Jillian, and Jon Henley. "European Investment Bank to Phase Out Fossil Fuel Financing." The Guardian, November 2019. https://www.theguardian.com/environment/2019/nov/15/european-investment-bank-to-phase-out-fossil-fuels-financing; Kottasová, Ivana, Ingrid Formanek, Angela Dewan, and Rachel Ramirez. "'Historic Breakthrough': 20 Countries Say They Will Stop Funding Fossil Fuel Projects Abroad." CNN, November 2021. https://www.cnn.com/2021/11/03/world/countries-agree-to-end-fossil-fuel-financing-abroad-cop26-climate/index.html.

43. CDP. "New Report Shows Just 100 Companies Are Source of over 70% of Emissions." July 2017. https://www.cdp.net/en/articles/media/new-report-shows-just-100-companies-are-source-of-over-70-of-emissions; Axelrod, Joshua.

"Corporate Honesty and Climate Change: Time to Own Up and Act." NRDC, February 2019. https://www.nrdc.org/experts/josh-axelrod/corporate-hone sty-and-climate-change-time-own-and-act; Science Based Targets. "Companies Taking Action." 2022. https://sciencebasedtargets.org/companies-taking-act ion; RE100. "RE100 2021 Annual Disclosure Report." 2021. https://www.there 100.org/; Eavis, Peter, and Clifford Krauss. "What's Really Behind Corporate Promises on Climate Change." New York Times, May 2021. https://www.nyti mes.com/2021/02/22/business/energy-environment/corporations-climate-cha nge.html.

44. Project Drawdown, "The Drawdown Review"
45. UNICEF. "The Climate Crisis."
46. Project Drawdown, "The Drawdown Review"; World Bank Group. "The Global Health Cost of Ambient PM2.5 Air Pollution." 2020. https://documents1.worldb ank.org/curated/en/202401605153894060/World-The-Global-Cost-of-Ambient-PM2-5-Air-Pollution.pdf.
47. Dennis, Brady, and Adam Taylor. "People Around the World Increasingly See Climate Change as a Personal Threat, New Poll Finds." Washington Post, September 2021. https://www.washingtonpost.com/climate-environment/2021/ 09/14/climate-change-threat/.
48. Sierra Club. "Yes, Actually, Individual Responsibility Is Essential to Solving the Climate Crisis." November 26, 2019. https://www.sierraclub.org/sierra/yes-actua lly-individual-responsibility-essential-solving-climate-crisis.
49. Baker, K. C. "Earth Day 2020: What You Need to Know and Do to Save the Planet—Before It's Too Late." People, April 2020. https://people.com/human-interest/earth-day-2020-what-you-need-to-know-and-do-to-save-the-planet-bef ore-its-too-late/.
50. Goldstein, Benjamin, Dimitrios Gounaridis, and Joshua Newell. "The Carbon Footprint of Household Energy Use in the United States." Proc Nat Acad Sci 117, 32 (2020): 19122–19130. https://doi.org/10.1073/pnas.1922205117.
51. World Resources Institute. "Shifting the Diets of High Consumers of Animal-Based Foods Could Significantly Reduce Per Person Agricultural Land Use and GHG Emissions." April 20, 2016. https://www.wri.org/data/shifting-diets-high-consumers-animal-based-foods-could-significantly-reduce-person.
52. Imperial College London. "9 Things You Can Do About Climate Change." 2021. https://www.imperial.ac.uk/stories/climate-action/.
53. Graziano, Marcello, and Kenneth Gillingham. "Spatial Patterns of Solar Photovoltaic System Adoption: The Influence of Neighbors and the Built Environment." J Econ Geography 15, 4 (2014): 815–839. https://doi.org/ 10.1093/jeg/lbu036; Westlake, Steve. "A Counter-Narrative to Carbon Supremacy: Do Leaders Who Give Up Flying Because of Climate Change Influence the Attitudes and Behaviour of Others?" October 2, 2017. https://ssrn. com/abstract=3283157.
54. Imperial College London, "9 Things You Can Do."
55. Baker, "Earth Day 2020."
56. Chakradhar, Shraddha. "New Pediatrician Network Puts Spotlight on Climate Change's Effects on Children." December 18, 2020. https://www.statnews.com/ 2020/12/18/new-pediatrician-network-puts-spotlight-on-climate-changes-effe cts-on-children/.
57. KC Baker, "Earth Day 2020: What You Need to Know and Do to Save the Planet—Before It's Too Late". https://people.com/human-interest/earth-day-2020-what-you-need-to-know-and-do-to-save-the-planet-before-its-too-late/.

INDEX

For the benefit of digital users, indexed terms that span two pages (e.g., 52–53) may, on occasion, appear on only one of those pages.

Tables, figures, and boxes are indicated by an italic *t*, *f*, and *b* following the page/paragraph number.

Heschel Center, 121–22
Hirsi, Isra, 133
Historically Black College and University
    Climate Change, 106
Houska-Zhaabowekwe, Tara, 112–13
Housty, Jess, 109
Hurricane Harvey, 62
Hurricane Ida, 15–16, 62
Hurricane Katrina, 62, 70, 72, 87
hurricanes, 14, 15, 61–62, 86–87
Hurricane Sandy, 62, 162
hydraulic fracturing (fracking), 6–7
hydrocarbons, 64–65
hydrofluorocarbons (HFCs), 170
hydrogen fuel, 177–78

Ibrahim, Hindou Oumarou, 109–10
ice-albedo feedback, 20
immune defenses, 34, 38, 50, 97–98
India, 5, 49–50, 63, 82, 83f, 88
Indigenous activism, 109–13
Indigenous knowledge, 18, 81–82
Indigenous land rights, 175–76
Indigenous people, 89, 90
industrial waste prevention, 170–71
inequality, 3, 41, 113–14, 167–68, 179.
    see also disparities
infants
    air pollution deaths, 50
    biologic defenses, 33–34
    heat-related deaths, 63–64
    heat vulnerability, 34–35
    mortality due to air pollution
        exposure, 84
    mortality due to heat, 63–64, 90
    nutritional needs, 35–36
    preterm, 58–59
    susceptibility factors, 24–25
    toxic exposure brain effects, 99
    toxic exposures, 29–33
    toxic interference, 36–38
infection, infectious disease, 34, 36, 58,
    68–69, 85, 92
inflammation, 38, 97–98
information processing networks, 26
Inhofe, James, 114
innate immunity, 34
Insua, Tomás, 115
Interfaith Center for Sustainable
    Development, 123

Interfaith Power and Light, 123–24
intergenerational equity, 105
intergenerational injustice
    defined, 81–82
Intergovernmental Panel on Climate
    Change (ICOG), 3
International Panel on Climate Change
    (IPCC), 124
IQ loss, 36, 51, 52–53, 56, 76,
    78–79, 99
Islam, 119–20
Islamic Declaration on Global Climate
    Change, 119–20
Israel, 15–16
Israel Climate Forum, 121

Japan, 5–6, 15–16, 53
Jedrychowski, Wieslaw, 148–50, 151
Jernigan, Terry, 28
Jewish Climate Action Network, 121
Jewish Climate Change Campaign, 121
Jewish Women in Environmental
    Activism, 121
Jilani, Hina, 135–36
Johnson, Boris, 143–44
Jordan, 15–16, 71
Judaism, 120–22
Juliana v. United States, 130, 183

Kahn, Sadiq, 143, 144–48
Klein, Naomi, 3, 109, 130–32
Kolbert, Elizabeth, 18–19
Koli, Zainab, 120
Krakow, Poland, 153–54 n.27
Kurdi, Alan, 92–93

Lancet, 46–47, 49–50
Lancet Commission Reports, 75–76, 125
landfill siting, 106
Landrigan, Philip, 24 n.1
Laudato si' encyclical, 113–16, 123
Lausanne Movement, 116
lead poisoning, 27–28, 36, 48 n.3, 82,
    92, 111–12
learning disabilities, 85–86
Lebanon, 71
LED lighting, 169–70
Licker, Rachel, 187
Lin, Maya, 179
Livingstone, Ken, 143